Innovation in Techno
and Organization

Innovation in Technology and Organization

Peter Clark and Neil Staunton

Foreword
by Everett Rogers

R

Routledge
London and New York

First published 1989 by Routledge
11 New Fetter Lane, London EC4P 4EE
29 West 35th Street, New York, NY 10001

Reprinted 1990
New in paperback 1993

Phototypeset in 10pt Times by
Mews Photosetting, Beckenham, Kent
Printed in England by Clays Ltd, St Ives plc

British Library Cataloguing in Publication Data

Clark, Peter, *1938–*
    Innovation in technology and organization.
    1. Innovation
    I. Title, II. Staunton, Neil, *1952–*
    303.4'84

    ISBN 0–415–00422–5 (Hbk)
    ISBN 0–415–09318–X (Pbk)

Library of Congress Cataloging in Publication Data

Clark, Peter A.
    Innovation in technology and organization / Peter Clark and Neil
Staunton.
        p.   cm.
    Bibliography: p.
    Includes index.
    ISBN 0–415–00422–5
    Technological innovations. 2. Organizational change.
3. Organizational behaviour. I. Staunton, Neil. 1952–
II. Title.
T173.8.C5 1989
658.5'1–dc20   89–33188
                                                        CIP

Contents

Contents

Contents

Figures and tables

Figures and tables

Foreword

Research on the diffusion of innovations began about 50 years ago as studies of individual decisions to adopt a new idea: Hybrid seed corn by Iowa farmers, a new antibiotic drug among medical doctors, and the decision to adopt 'modern math' by school teachers. Only much later did scholars begin to investigate innovation in organizations, despite the fact that much of our lives, especially our work lives, is spent in organizations. Thus students of innovation came together with scholars of organizational behaviour to create a hybrid scholarly field that is distinctive from either parent. The offspring is showing a good deal of hybrid vigour these days.

The present book is written by one of the leaders in this new type of organizational studies of innovation, and he brings out the best of the parental research traditions in the new generation. Dr Clark conducts investigations on innovation in organization, and this close relationship with the topic shows in the skilful way in which the book is organized and written. The authors also draw upon the growing body of studies of this genre conducted by other scholars in Europe and America. They provide a useful framework in which the empirical findings are placed. The book's graphics, which help illustrate the framework, are excellent and add to the readability of the volume.

For example, a typology of five levels of innovations is provided, from generic, to epochal, to altering, to entrenching, to incremental. This classification of the degree of change represented by an innovation is an important addition to the frameworks utilized in past research. Certainly the degree to which a new idea implies radical versus incremental change is a key dimension. Unfortunately, past studies have treated all innovations (from miniskirts to snowmobiles to computers) as essentially similar. Obviously they are not.

Dr Clark and Neil Staunton explain that much of the current policy emphasis upon innovation stems from Japanese industrial competitiveness in technological innovation. While Japan has not taken the lead in creating innovations, it has excelled in recent decades in transferring the

Foreword

technological ideas from their R&D origins into innovative products. As American and European companies have increasingly lost their market share in recent decades, they (and their governments) have sought to better understand the innovation process and to influence this process through various policies. One outcome has been the increased number of studies of innovation in organizations.

In this book, the authors utilize a theoretical repertoire of concepts including re-invention, exnovation, longwaves, centre-periphery, population ecology, structuration, and transaction costs. Case study analyses are utilized here to illustrate and integrate this theoretical framework. The cases include McDonald's (the company that sells one-third of all the hamburgers in the United States), Rover automobiles, Marks and Spencer, and Rugby Union football.

The overall result is a book that is readable, comprehensive, and theoretically integrated. I commend it to your attention.

Everett M. Rogers
Annenberg School of Communications
University of Southern California

Preface/acknowledgements

This book bridges two research programmes at the Aston Business School in Birmingham, England, undertaken between 1982 and 1990. The first was funded by the Economic and Social Research Council with a programme grant to Peter Clark, Jennifer Tann, John Child, and Ray Loveridge at the Work Organization Research Centre. This had the objective of investigating innovation in technology and organization in eight sectors of the British economy over a period of two decades. The current programme is from the Science and Engineering Research Council and consists of a portfolio of four projects in the Innovation, Design, and Operations Management group. Our specific project concerns the study of the factors affecting the choice, design, and implementation of computer-aided production systems, and involves David Bennett, Peter Burcher, Adrian Campbell, Peter Clark, Sudi Sharifi, Neil Staunton, and Jennifer Tann. Whilst the former programme had a design orientation to innovation, the latter takes a diffusion orientation. Their blending has stimulated our approach and demonstrated its relevance to the general shift which is occurring in organization studies that we reported in *Organisation Transitions and Innovation Design* (Peter Clark and Ken Starkey, Pinter & Columbia University Press, 1988).

Our colleagues under the Work Organization Research Centre umbrella in the Technology Policy Unit and in the Social Innovation Research Group have provided a positive and constructive ambience. Outside Aston we have benefited from Paul Stoneman (Warwick) and Rod Coombs (University of Manchester Institute of Science and Technology) on the economics of technical change and innovation. Alf Thwaites, John Goddard, Neil Alderman and Chris Pywell at the Centre for Urban and Regional Studies, Newcastle University, provided an invaluable commentary on our early thinking.

Our study of diffusion has been greatly assisted by the advice of Gerry Waterlow, consultant to the Science and Engineering Research Council, who has overall responsibility for the £2 million initiative to promote the use of computer-aided production management. Their pilot study of computer-aided production management was a model of careful enquiry.

Due to an assortment of locations being involved in the writing of this book, five word-processing systems had to be learnt and their contents conveyed to Aston. Many thanks to Joy Simkin and Lesley Baker for their assistance with this.

Jane Winder and Patricia Clark, at the Innovation, Design, and Operations Management group, provided the modern form of secretarial support to the production of this manuscript: subjecting our spelling to the lexicon, removing some of our innovative new vocabulary, locating references, pacing our writing and revisions, and advising on new developments for graphics, which were excellently carried out by Debbie Evans. We are highly appreciative.

We are conscious that certain thinkers formed the axes around which our interpretation has unfolded. We want to acknowledge our special indebtedness to Everett Rogers and to the late Bill Abernathy. Our thinking has always been stimulated by the local dialectics and its key luminaries whom we heartily praise. The responsibility for interpreting these many suggestions is certainly ours. Our intention is to provide a relevant and stimulating pathway into this pluri-paradigmatic realm.

Finally, our thanks go to Andrew Lockett of Routledge, without whose support and encouragement this book would never have come into being.

Peter Clark and Neil Staunton
Aston Business School,
Birmingham, England

Part I

The new perspective

Chapter one

Innovation: limits to the orthodox mainstream

They [economists] accept the data of the momentary situation as if there were no past or future to it . . . the problem that is usually being visualised is how capitalism administers existing structures, *whereas the relevant problem is how it creates and destroys them.*
Schumpeter, *Can Capitalism Survive*, chapter 7, 'Creative destruction'

Introduction

All organizations face the dilemma of balancing and blending their relative orientation to efficiency and to innovation (Abernathy 1978; Lawrence and Dyer 1983; Clark and Starkey 1988). For much of the twentieth century there has been a tendency to give efficiency a priority over innovation, but this tendency began to alter in the 1970s. Today innovation is surpassing efficiency as the primary principle for deciding the most appropriate form of organization.

Although all organizations are affected by the new significance of innovation, the degree of impact differs because enterprises are being divided into two broad classes: those which are focal centres of innovative synthesis for a network of other enterprises (for example, McDonald's, Marks and Spencer, Toyota, Benetton); and those who are intermediate suppliers within a network. The problems of balancing efficiency with innovation arise for the focal enterprises and for the intermediate suppliers, yet the requirements for blending innovation with efficiency have altered. That shift in balance presents enterprises and their academic analysts with the limits to existing styles of analysis. This book seeks to map the problems and to identify the most fruitful lines – theoretical, practical, and empirical – which will be central to future development. Our point of entry into the examination of innovation in technology and organization is mainly, though not only, from an organization perspective. Organization became a major focus of knowledge–building three decades ago when several orientations and theoretical perspectives were produced in an attempt to prise open the lid of the 'black box' of the

3

firm (for example, March and Simon 1958; Woodward 1958; Penrose 1959; Burns and Stalker 1961). An important feature of those early theories was their synthesis across existing disciplines. Since then various new disciplinary constellations have been articulated, have crystallized, and have become rather rigid. This feature is particularly so for what has become known as the 'Organization Sciences' or 'Macro Organization Behaviour' in the USA, and known as 'Organization Studies' in Europe.

It is our contention that the problem of innovation in macro organization behaviour has been theorized in a very restrictive and simplistic manner which is now quite unsatisfactory. However, we do not propose to abandon the entire analytic inheritance. Instead, we propose that the fulcrum of theory building and policy analysis is shifted from an implicit focus upon efficiency, with innovation as the deviant case, to innovation as the crucial focus, with efficiency as a necessary adjunct (see chapter 2). The role of innovation is mediated by design (see chapter 10). This revision in the orthodox mainstream is necessary in order to provide the kinds of analysis which are relevant to the pressing problems of adaptation in contemporary enterprises.

It is useful to consider whether the growing attention to innovation has arisen from the context or from within the various disciplines focusing on innovation. It may be argued that the growth of interest in innovation has very largely been triggered and stimulated by events in the contexts of organizations. There is widespread agreement that the past two decades have witnessed the demise of many well-known enterprises and the emergence of many new enterprises, especially from the Pacific Basin. Of course many old, large enterprises have survived. Also, a good number of previously medium-sized enterprises have become much larger, and have emerged as the shapers of complex, multiple chains of other enterprises. Explaining this new situation is a matter of debate and there are several contending theories.

Describing these transitions in the contexts of enterprises and locating the new forms of best practice in technology and organization has revealed important limits in the existing stock of knowledge known as macro organization behaviour. The same problem also besets the parallel areas such as the economics of technical change. The purpose of this opening chapter is to sketch the new agenda and to explain the syntax of the book. This chapter juxtaposes two areas: the changing contexts of enterprises and the degree to which the analysis of the changing contexts could be satisfactorily undertaken within established, orthodox approaches. We argue that orthodox approaches have been too strongly committed to efficiency. Consequently the orthodox theories contain a series of limiting restrictions and contradictions. We propose that the appropriate vision for theory-building in macro organization behaviour should be innovation rather than efficiency. After introducing these issues the final section of this chapter overviews the structure of the book, explaining how the new mainstream may be developed into a useful theory of innovation.

Explaining the current contexts

The core agenda of macro organization behaviour was crystallized in the early 1960s around the issues of structure and contingency in the form of equilibrium and functional models. The knowledge created by this process was intended to provide a 'hard' framework for evaluating existing forms of structure and for designing alternatives rather than as an explanatory-descriptive form of knowledge (Clark and Starkey 1988: chapter 1). Consequently, macro organization behaviour has played a relatively small part in the analysis of the contextual changes which arose during the 1970s and 1980s and whose unfolding still continues. So the main impacts on macro organization behaviour have come from without rather than within.

The major external stimulus on macro organization behaviour has probably been the success of the Japanese in penetrating the North American markets through high sales in automobiles and electronic consumer goods to the point where serious imbalances of trade occurred between Japan and the USA. Initially the success of the Japanese was attributed to lower labour costs, but the detailed comparisons between the USA and Japan revealed two surprising features:

(a) The Japanese had much lower inventory costs because their leading firms practiced a form of just-in-time 'pull' system. This discovery triggered an interest in Just-in-Time and led to a plethora of books on achieving competitiveness through manufacturing.

(b) The Japanese invested heavily in American patents, yet had developed extensive skills in design and development rather than in research and development (R&D). It seemed that the benefits of American investment in the new knowledge for the invention stage was being appropriated by the Japanese skills in exploiting scale economies.

Examination of these two features has led to a much more critical and reflective scrutiny of American corporate practices. Also, analysts agree that American hegemony as the inventor and supplier of best practice on innovation in technology and organization is being challenged, but they differ on the reasons, on the directions, and on the consequences.

Various reasons are offered for the American loss of hegemony in defining the best practice organization and innovation. A brief considera-tion of five recent investigations reveals the complexity of interpreting current events. First, analogies have been drawn between the apparent decline of the USA and the earlier declines of major empires built in the Mediterranean world of seventeenth-century Spain and the more recent relative decline of Britain (Clark 1987). It has been argued that the USA is unable to bear the economic costs of its foreign policy (Kennedy 1987). Some argue that military expenditure has induced an uncompetitive collection of major enterprises. Second, some of the postulated strengths of American corporate activity have been shown to have been weaknesses. Johnson and Kaplan (1987) claim that managerial accounting was used much less stringently and very much less systemically by American corporations than had been presumed. Likewise, Abernathy, Clark, and Kantrow (1983) contend that firms relied on systems of control which wrongly buffered the core processes from external shocks (cf. Thompson 1967) and that the constitution of knowledge in the management sciences provided an illusory image of managerial control. These recent assessments suggest that Chandler's (1977) account of American management as the visible hand replacing the market omitted to analyse the distinctive features of the American market and to scrutinize the longer-term consequences. Also it has been argued that America's real strength was in distribution (for example, railways, telegraph, retail, mail order) and marketing rather than in manufacturing. Consequently, the previous assumption that America was best at large-scale production has been revised by Schonberger (1982) to suggest that the Americans are best at medium-batch production. Third, the former image of the USA as highly competent at design for innovation (Noble 1977; Pulos 1983) has been questioned by the new claim that the unfolding of the design cycle is too rigid because of the highly segmented, autonomous and adversarial relationships between departments within the techno-structure. Fourth, Piore and Sabel (1984) contend that for a century after 1870 the economy and society which was best 'fitted' to the hidden rules of the invisible hand of Darwinian selection criteria for economic accumulation was the USA. Since the 1970s the hidden rules have begun a period of transition comparable to an earlier transition between 1850 and 1870. They refer to that earlier period as the First Divide and the current period as the Second Divide. After the First Divide the basis for success was (they contend) mass production along so called Fordist lines. American firms and institutions were well adapted to these forms. The interpretation of Piore and Sabel suggests that after the Second Divide (1970–90) flexible specialization will become the main requirement for success, and they claim that this requires forms of organizing which are analogous to the community-anchored businesses of nineteenth-century cities like Birmingham in England and Lyon in France. The Second Divide thesis has been a useful focus for reinterpretations, but that thesis is in the process of substantial revision. Fifth, the revival of interest in long-wave theories and also in the neo-Schumpeterian theories of economic pulsations, and of the swarming of innovations, offers yet another possibility. If, as these long-wave theories suggest (see chapter 5), there are fifty-year pulsations containing equal periods of upward and downward movement, then it could be argued that during the period from 1940 to 1970 America maintained a dominant position during the upward phases when the new cluster of innovations were being commercially exploited through intensive research and development. That, after the mid-1960s further research and development becomes less beneficial and the key necessity became to shift to design and development and to economies of scale. It might be added, through borrowing from the Second Divide thesis, that the downswing coincided with a market shift in which the economies of scale (formerly an American strength) had to be combined with economies of scope (formerly a European strength). If this interpretation is to be followed, then the Japanese case is one of exploiting the downswing through blending scope and scale whilst investing research and development in the next cluster of innovations rather than on the outgoing cluster.

These five examples are sufficient to highlight the point that there are ongoing contextual shifts in markets and in the character of innovations whose detection, description, analysis and theorizing for policy purposes lies outside orthodox macro organization behaviour.

Limits

Orthodox macro organization behaviour did contain much more variety and disputing of perspectives than many recent accounts have suggested (see Clark and Starkey 1988), though there was only the slightest degree of the use of different perspectives to interpret the same phenomena (Eisenstadt 1973; Burrell and Morgan 1979; Lammers 1981). So it is worth noting two lines of enquiry which might have formed the 'seed' for germinating a more thorough going vision of innovation in technology and organization. First, the appearance of the best-selling analyses on the excellence theme provided an extraordinary, though unsynthesized, coupling of the highly complex Weickian perspectives with highly simplistic general traits which were reminiscent of the popular documents 'how to do it' pamphlets for American agricultural farmers in the mid-nineteenth century (Weick 1979; Peters and Waterman 1982). These traits, and the subsequent literatures on corporate renewal, have all proceeded in an attempt to 're-paradigm' American corporate managements. Second, Aldrich (1979) presented a bold and relevant prospectus for re-examining the diffusion of innovations and the problems of being innovative. Like the perspectives of Weick, the prospectus of Aldrich has been more honoured in the footnotes than in the shifts in the type of law-like knowledge which macro organization behaviour favours. So, we contend that the orthodox mainstream of macro organization behaviour has reached the limits of the capabilities of its main perspectives, and that a roster of serious problems can be identified. These problems are so pervasive that their puzzling features require an alternative vision to efficiency as the pivot of theory construction. We propose that innovation should be the pivot. The limits to orthodox macro organization behaviour have created an array of problem areas in the attention to, and treatment of, innovation. There are ten major problems as itemised below.

First, many studies have imposed an objectification on innovations so that an innovation is treated as a 'thing' which is detached from its contexts and pathways. Also the plurality of players (for example, suppliers, users) are ignored. Objectification leads to a limited and flawed understanding, yet it is the most widely adopted procedure of analysis.

Second, there is a strong tendency to equate innovations with equipment and to neglect the knowledge which is embodied in other dimensions such as raw materials, layouts, and in standard operating procedures. Too little attention has been given to the role of unembodied knowledges (see Karpik 1978), to the application of the knowledge technology perspective (for example, Perrow 1967) and to the fruitfulness of

Figure 1.1 Utterback–Abernathy framework

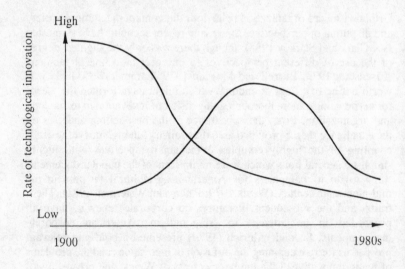

Source: Utterback, J.M. and Abernathy, W.J. (1975).

Figure 1.1 continued

	Fluid pattern	Transitional pattern	Specific pattern
Competitive emphasis on	Functional product performance	Product variation	Cost reduction
Innovation stimulated by	Information on users' needs & users' technical inputs	Opportunities created by expanding internal technical capacity	Pressure to reduce cost & improve quality
Predominant type of innovation	Frequent major changes in products	Major process changes required by rising volume	Incremental for product & process, with cumulative improvement in productivity & quality
Product line	Diverse, often including custom designs	Includes at least one product design stable enough to have significant production volume	Mostly undifferentiated standard products
Production processes	Flexible & inefficient; major changes easily accommodated	Becoming more rigid, with changes occurring in major steps	Efficient, capital-intensive, & rigid; cost of change is high
Equipment	General-purpose, requiring highly skilled labour	Some subprocesses automated, creating 'islands of automation'	Special-purpose, mostly automatic with labour tasks mainly monitoring & control
Materials	Inputs are limited to generally-available materials	Specialized materials may be demanded from some suppliers	Specialized materials will be demanded; if not available, vertical integration will be extensive
Plant	Small-scale, located near user or source of technology	General-purpose with specialized sections	Large-scale, highly specific to particular products
Organizational control is	Informal & entrepreneurial	Through liaison relationships, project & task groups	Through emphasis on structure, goals & rules

Source: Utterback, J.M. and Abernathy, W.J. (1975).

using notions like paradigms (for example, Dosi 1984). The problem of equating innovations only with equipment has been referred to as 'the problem with technology' (McDonald *et al.* 1983).

Third, for assuming that the occurrence of innovation is a detour which will be followed by a return to normalcy. This viewpoint has been common amongst managements and has sometimes led to regarding innovation as a leap ahead of rivals which is then followed by stability. It is important to recognize this perspective and to assess its prevalence amongst those most centrally connected with innovation: capital, management, and labour. The return to normalcy needs to be scrutinized. This is a difficult problem to untangle and it is not resolved by adopting a simplistic pro-innovation stance.

Fourth, there is a widespread tendency to conceptualize innovation as a sharp dichotomy between radical and incremental innovation. Often this is done unwittingly because the dichotomy has achieved the status of a stylized fact. One of the most pervasive examples is the Utterback–Abernathy (1975) framework which depicts the longitudinal patterning of radical and incremental innovation over the lifetime of the sector as shown in Figure 1.1.

The assumption is that at the founding of the sector there is radical innovation in the product which is followed by radical innovation in the production process and then, in due course, there is widespread incremental innovation. The Utterback–Abernathy framework has been coupled with the biological metaphor of the life cycle from birth to maturity. These and similar frameworks are highly popular, yet practically disastrous and require replacement by more discriminated typologies (Abernathy, Clark, and Kantrow 1983).

In order to establish a more refined analysis of degrees of innovation we introduce a simple set of distinctions which will be anchored and refined in the following chapters. Five levels are characterized, commencing with the most extensive and radical forms:

1. generic innovations which create new techno-paradigms (Freeman 1982; Perez 1983) of clusters of innovations emanating from a new core process (for example, steam engines, electricity, internal combustion engines) which cross-cuts many sectors and many stages of production. Generic innovations frequently outpace the capacities of existing institutional networks to adapt and to incorporate them. The point is clearly illustrated for urban electric generation during the earlier part of this century when there were sharp differences between Chicago, Berlin, and London in their capacity to create networks of power (Hughes 1983). Consequently, new techno-paradigms may fit most easily in social structures which are quite different to the forms leading states or enterprises. Systemofacture

(see Kaplinsky 1984), for example, seems to fit very easily into the Japanese enterprise and society where there is slight segmentation of managerial work, whilst in the USA, where there is extensive segmentation of managerial tasks, there are many problems of incorporating systemofacture. The early stages of a new techno-paradigm are likely to create crises of institutional regulation in the struggle between existing power centres and those perceiving new opportunities for power. Generic innovations may occur in bunches and their unfolding may possess a long-wave pattern (see chapter 5);

2. epochal innovations are subsets of generic innovations of considerable magnitude whose introduction (or rejection) is confined to particular sectors of activity. Examples would include the introduction of the catalytic cracking process into the chemical industry, and the development of automatic gear-change in the automobile industry, or the development of specific items like Plexiglass and Kodachrome. These may lead to the re-founding of existing sectors (see chapter 5);

3. altering innovations which introduce important alterations at the level of the firm. Examples would include electronic point-of-sale systems in retail which, although entrenching certain ongoing directions, also enable system-wide alterations in the connectedness of activities and in the location of initiatives;

4. entrenching innovations which modify existing methods, but proceed in the same direction. These often require considerable detailed adjustment on an incremental basis which cumulates into significant developments;

5. incremental innovation in which there are no new inputs, but the existing collection of inputs are reconfigured to achieve a higher output from the system.

The typology of generic, epochal, altering, entrenching, and incremental innovation provides an initial set of distinctions which can be refined. The tendency to treat innovation only as a dramatic event is common in the organization sciences.

Fifth, there is strong persistent tendency to treat innovation as a neutral process which is only guided by processes such as 'induced innovation' (for example, Rosenberg 1976,1982). This approach deliberately ignores the impacts on the existence and form of innovation in any setting because the objective is a law-like general theory. However, such theories are only useful to a very limited extent because they obscure the detailed examination of innovation as a process and because they confine policy initiatives. What is needed – if the 'black box' of the firm is to be prised open – is a perspective which consciously examines the role of social structures and actors, especially the struggles which accompany and impact the paths which innovations take. Latour (1986) observes that the opening of the frequently invoked 'black box' might well reveal that it was populated by 'Sartrian engineers' whose regimes of problem solving were rather different to those forms suggested in the rationalistic literature on innovation. Many innovations are unintended outcomes of such struggles (Clark 1987). Moreover, the reference to concepts such as 'bottle-necks' requires the reconstruction of the mentalities of managers (for example, American) who perceive the world in terms of flows compared to those who do not. The economics of technical change has suffered greatly from the depiction of American practices as a neutral process. Equally limiting is the overstructured conception of processes and the neglect of individual biographies and political struggles. This feature is very evident in the examination of top biographies and their role in the corporate restructuring around an epochal innovation in a large mail-order firm (Pettigrew 1973).

Sixth, there has been a strong bias towards innovation which has led to an almost total neglect of the problems of the removal of existing practices so that they can be replaced: exnovation. This deficiency is particularly strong in macro organization behaviour (Kimberley 1981). It is therefore important to discover how the analysis of exnovation can be improved. For example, the notion of creative destruction is very powerful at the macrosocietal level (Schumpeter 1939), and it may be argued that Thatcherist Britain has been a prime example of where the state has intervened to disrupt long-established practices of occupational regulation at many levels. In Britain these same principles of exnovation are operating within firms through a variety of devices: closures of plants, sacking and ablation (for example, universities), take-overs by finance capital, offloading existing functions into a competitive and subcontracting mode (for example, hospitals). Yet the important issue is to examine how the coupling of exnovation and innovation occurs (Yin 1979).

Seventh, too little attention has been given to examining the specific market contexts in which enterprises have elected to operate. It may be postulated that markets exert a persistent, almost hidden, process of entraining organizations to their basic patterns and structure (Weick 1969; Clark 1987). So, given that strategic decision-makers 'choose' their markets, then more attention should be given to examining how the characteristic features of different markets may constrain or enable the 'learning paths' of innovation. There is already a significant amount of analysis which confirms the relevance of this perspective, so the problem is to contextualize the analysis of contemporary innovation so that its interfaces and embedding can be characterized. This requirement is especially important when considering the transfer of innovations between countries. For example, recent analysis suggests that the Japanese consumer market is as homogenized as that of the USA, whilst also containing a heavy concentration in Tokyo of consumers who have high disposable incomes and high expectations from sophisticated electronic household goods (for example, white goods, 'smart systems', video cameras). Also it is suggested that in Japan the speed of market response to new products is very high so that successful products could travel down the learning curve from small to medium and larger batch size at a great speed. These points may be compared with the markets chosen by British firms. The notion of socially-enacted entrainment to the market may be applied to these features (Clark 1989).

Eighth, there is extensive Balkanization of problems between the various disciplines which address the general area of innovation: economics of technical change, management of technology, macro organization behaviour, industrial organization, spatial geography. Even within the economics of technical change there are several disjoint intellectual traditions so that there are no conceptual bridges between the major exemplars: Sappho, Hindsight, Wealth from Knowledge, Poete, the approach of Grilliches, and that of Mansfield. In practice several attempts at synthesis are underway, most obviously in the pooling of knowledge about the diffusion of innovations amongst economists since the Venice Symposium of 1986. Yet these massive attempts at synthesis will and do proceed rather slowly. The aim of this book is to provide a constructive vision of synthesis centred on macro organization behaviour and, where feasible, to make bridges to other disciplines so that forays of joint disciplinary work can become more focused and more fruitful. One of the major requirements is for joint research projects into the detailed empirical analysis of innovation in technology and organization. There is a distinct lack of such detailed studies (Nelson and Winter 1977: 41).

Ninth, there is also extensive Balkanization between problems which ought to be examined in a connective manner. The most obvious instances are the linear scheme favoured by economists of technical change (that is, invention/commercial innovation/diffusion) and the stage model of designing/implementation/and consequences. The latter separation is very awkward. For example, a great deal of current attention is being given to the implementation of innovations, most obviously by those involved in the 'management of technology'. However, this results in the neglect of design as a cycle of activities which embraces the implementation and which seeks to anticipate the consequences (Clark 1972a, b; Whipp and Clark, 1986). Amongst those in organization studies there has been a tendency to omit the design state, to give limited attention to implementation, and to decontextualize the consequences.

Tenth, there is a strong tendency to use atemporal, cross-sectional models, or to impose an overly smooth pattern of transitions in state by the specific enterprise and by populations of enterprises (Clark and

13

Starkey 1988). Transitions by enterprises of a major kind are highly problematic (Miller and Friesen 1984) and too little is known about the organizations which were not turned around. What is needed are sample studies comparing survivors and non-survivors.

These ten problem areas with orthodox macro organization behaviour arise because existing perspectives are not rich enough to illuminate the contemporary events in a relevant fashion for policy-making and for undertaking strategic innovation. Existing perspectives assume the vertically integrated, structurally homogeneous enterprise which is controlled from the top and steered by the Balkanized expertise of the techno-structure. We have reached the limits to the usefulness of existing macro organization behaviour perspectives in their current mode. A new mode, and a more relevant set of underlying metaphors, are required. The previous mode was articulated and bonded through March and Simon's (1958) usage of the computer and its programs. That certainly was an advance on analogies such as the mechanical machine, yet what is required is a fuller understanding of organisation in its own right and that needs the replacement of many existing metaphors such as the life cycle. That said, there are important analytic principles from orthodox macro organization behaviour which can be carried forward and integrated into the new mainstream (see chapter 2). Amongst these analytic principles which are still required is a broader and deeper notion of the information costs perspectives associated with Perrow's perspective on social technology (Perrow 1967; Galbraith 1977). However, these analytic principles have to be 're-set' within a vision of innovation as the core, strategic problem for survival and success.

Structure of the book

The aim of *Innovation in Technology and Organization* is to construct a synthesis on innovation from the position of macro organization behaviour. Our aim complements similar, parallel endeavours in the economics of innovation (for example, Clark, N.G. 1985; Coombs, Saviotti, and Walsh 1987) and in the management of technology (for example, Rhodes and Wield 1985; Roy and Wield 1986) and in marketing (for example, Foxall 1984). Where relevant we attempt to synthesize from these complementary perspectives.

Economists contend that innovation can be viewed as a more or less linear process of three stages: invention, commercial innovation (prototypes into production), and the diffusion of innovations, but macro organization behaviour has given very slight attention to invention and to the role of research and development (cf. Freeman 1974; Combs *et al.* 1987). Macro organization behaviour has relied on two pathways for examining innovation, so it is relevant to see how our book utilizes these established routes: (a) innovation-diffusion; (b) innovation-design as corporate culture.

First, the diffusion of innovations has been a major area of enquiry in its own right, especially in the USA. The diffusion of innovations has revolved around the guiding themes articulated by Rogers (1962, 1983) and now known as the centre-periphery model of diffusion (Schon 1971). Interestingly, Rogers and Rogers (1976: chapter 6) made a significant foray into the processes by which organizations select and ingest externally available innovations, thereby providing a potential linkage with the approach of Burns and Stalker. That foray seems to have been suspended, yet Rogers and Rogers (1976: chapter 6) provided an outline sketch which requires further elaboration. That was an important attempt to examine the crucial problems of innovation diffusion and their appropriation (Clark 1987).The diffusion of innovations includes their international transfer between enterprises from different cultures as well as the internal transfer of innovations within the same enterprise in a single culture. The significance of this area has greatly increased with the recognition that the pool of solutions for the problems of organizing which has been developed over the past century is itself an impediment to contemporary adaptation (cf. Piore and Sabel 1984). Second, corporate innovation as a routine problem was solved by March and Simon (1958) and by Burns and Stalker (1961). Much contemporary writing on innovation is a very partial restatement of Burns and Stalker. An obvious example is the persuasively-written analysis of the adaptation of contemporary American corporations (for example, General Motors) by Kanter (1984) in which the organic systems solution is highly visible, but the problems of its usage are understated (cf. Burns and Stalker 1961: chapter 6). Likewise the model of the ambidextrous organization with collateral structures very much simplifies the political problems of innovation (Zaltman and Duncan 1977). A careful reading of Burns and Stalker reveals that their thesis was based on empirical examples of small-scale innovation in the branch plants of largeish corporations in textiles and electronics (Ghoshal and Bartlett 1987). There is very little on the undertaking of the full cycle of strategic innovation (cf. Whipp and Clark 1986). Indeed, as the revised preface to Burns and Stalker makes clear, their approach was highly static and too mechanical! So, this cherished, frequently invoked, analytic solution to the problems of organizational innovation requires close scrutiny.

We have 'chunked' the problems of innovation in technology and organization into four parts:

PART I: THE NEW PERSPECTIVE

PART II: INNOVATION

PART III: INNOVATION–DIFFUSION
PART IV: INNOVATION–DESIGN

These four parts are overlapping and interrelated:

Part I indicates how the subject matter of innovation in technology and organization requires refocusing (chapter 1) and presents a framework for examining the strengths and limitations of different perspectives on innovation (chapter 2). The proposed perspective is a general orientation containing several perspectives which share a focus on innovation as an uneven, structurally impacted, array of processes moving at different paces. Chapter 2 establishes the guiding principles and situates a series of examples from contemporary studies in terms of their success in serving these principles.

Part II applies the guiding principles to replace the objectification of innovation with an innovation configuration (chapter 3) in which unembodied knowledges (in the plural) are central and their various forms of embodiment in equipment, in layouts, in stabilized practices, and in raw materials are shown in their dynamic relationships. The innovation configuration is relational and its perception varies amongst vested interests such that its apprehension – the shapes – and the usages are problematic to varying degrees. The innovation configuration is therefore contingent and specific. Chapter 4 examines some key dimensions of the configuration. Particular attention is given to developing the distinction innovations which actually entrench pre-existing relationships from innovations whose usage requires and triggers (at varying paces) alterations in existing relationships. Too much previous analysis has either neglected this aspect totally or has assumed that all innovations are altering when many innovations are entrenching. Another key dimension is the extent to which generic and epochal innovations are substituting for existing innovations. This is an analytic problem which requires judgement and care. The possibilities are illustrated by reference to the 'sailing-ship effect'. Lastly some attention is given to the issue of whether innovations largely occur through deliberate strategic innovation in existing enterprises or through shifts in the composition of populations of organizations: the population ecology debate. Part II illustrates these themes with the case illustration of the American fast food industry in its founding period.

Part III 'chunks' the collection of issues which have been labelled innovation-diffusion into three chapters (5, 6, and 7). This part is very much concerned with developing and eventually replacing the orthodox perspective based on the centre-periphery situation with a selection of analytic elements which are portable and flexible. Part III commences with the establishment of the long-term dimension (chapter 5). To develop the essential long-term perspective and to reveal the impacted, uneven nature of macro processes we examine two frequently-cited models from the economics of innovation: the long-wave thesis and the Utterback–Abernathy model of sectoral development. Until recently these perspectives have been marginal and peripheral to macro organization behaviour. The long-wave thesis has certain heuristic merits, particularly in the models proposed by Freeman and by Perez. These may be set beside the Utterback–Abernathy framework to show why its revision is essential. Longitudinal analysis reveals that innovations are ensembles (Gille 1978) which flow down 'paradigmatic pathways', guided and impacted by human agency, containing chance and unintended outcomes as well as plentiful blind alleys which are only apparent *ex ante*. Chapter 6 examines the supply side of innovations starting with the cost-size corridor framework and closely scrutinizing the classic frameworks articulated in the earlier studies by Rogers (1962). Particular attention is given to the role of the suppliers in making innovations available – to suppliers' problems of retaining the profits and other benefits from developments of the innovation. Chapter 7 deals with the international transfer of innovations. This aspect is illustrated by reference to Anglo-American diffusion and this case is treated as the exemplar for examining transatlantic and transpacific (for example, Japan/USA) transfers of innovations. The Anglo-American case is a classic because macro organization behaviour has been responsible for conflating the differences and for obscuring the problems of transfers. The case of teamwork and its different meanings in the American and British cultures is utilized to illustrate the key issues. Brief reference is made to American influences in France, Germany, and Japan to highlight the directions for exploring the Japanization of innovation.

Part IV addresses the question of whether an enterprise possesses a cultural capability for engaging in the combination of innovation and design: innovation-design. This capability is normally located both within specific enterprises and in the network of which they are an aspect. Interfirm networks vary in the degree to which innovation is orchestrated from focal poles into chains of firms (chapter 8). An era when new clusters of innovation configurations are emerging poses challenges to existing interfirm networks – and to their poles – as well as providing the opportunity for new networks to be established. The establishment of the existing poles has often been a lengthy process interrupted by short, infrequent, unpredictable bursts of activity. The creation of new networks is also deceptive and pregnant with 'surprise events' like the coupling of firms founded in different long waves (for example, cars with aeroplanes and computer firms). Chapters 9 and 10 examine two core processes of innovation at the level of the organization (of all kinds): the ingesting of externally originated configurations and the design cycles for internally generated variants of specific goods and services. We propose a framework which situates strategic innovation in the context of the firm specific knowledge and the firm specific structural repertoire (in chapter 9) to show how macro organization behaviour should address the problems of innovation. Then the core problems of ingestion and the generation cycle are examined (in chapter 10).

These four parts are simply 'chunks' of the overall perspective. The application of an 'exquisitely innovation perspective' requires their compacting and focused usage.

Chapter two

New mainstream

Introduction

This chapter locates and explains how the new mainstream relates to the orthodox mainstream. A simple framework has been constructed to highlight the directions of our analysis. The framework is constructed from two core dimensions, each of which is split, to create four quadrants. This chapter introduces the two core dimensions and the four quadrants. The framework is heuristic and the two dimensions are composites. The framework is shown in Figure 2.1.

The new mainstream is located in the bottom right-hand quadrant and the orthodox mainstream situated in the top left-hand quadrant. The arrows from the orthodox mainstream indicate recent developments which both signify problems within the mainstream and also represent serious attempts to move towards the position we have termed innovation-design.

The purpose of the framework is to highlight the significance of developing an innovation perspective.

Framework: two dimensions

Efficiency versus innovation-design

The horizontal dimension distinguishes between two orientations in theory building and in practice:

(a)   towards efficiency as the organizing principle. This orientation is found in many current equilibrium models and in the functionalist models used in organization theory;

versus

(b)   towards innovation-design as the organizing principle with efficiency considered as a significant subset (Clark and Starkey 1988: chapter 4).

Figure 2.1 Core framework and quadrants

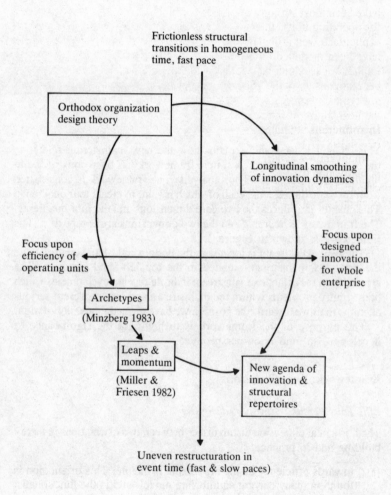

The orientation to efficiency is well established and therefore attention will be given to introducing the orientation to innovation.

Efficiency has been a prime concern of organization design theories (Galbraith 1977; Child 1984). Economics and the organization sciences have been most strongly oriented towards the prescribing of efficiency for operating units. In economics the orientation towards efficiency is explicit and well articulated, whilst in the organization sciences, which have been modelled on economics, the attention to efficiency is often implicit. The significance of the attention given to efficiency may be recognized in the time frame of the equilibrium models which are typically confined to the short run of the immediate present and the near future of the next financial quarter. Consequently the longer-term future is conflated and the degree of uncertainty is oversimplified (Gold 1981).

Innovation was largely assumed until very recently. The problems of innovation have been less central to organization studies (Clark 1987: chapter 1). Little attention has been given to the analysis of the capability of an enterprise as a whole to engage in innovation, especially with respect to the role of design in achieving adaptation (Clark 1987: chapter 5).

Figure 2.2 Designing processes

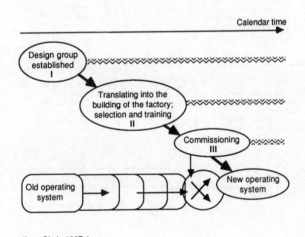

(from Clark, 1987a)

Source: Clark and Starkey (1988).

The new perspective

Figure 2.3 Strategic design – Rover SD1 (1968–82)

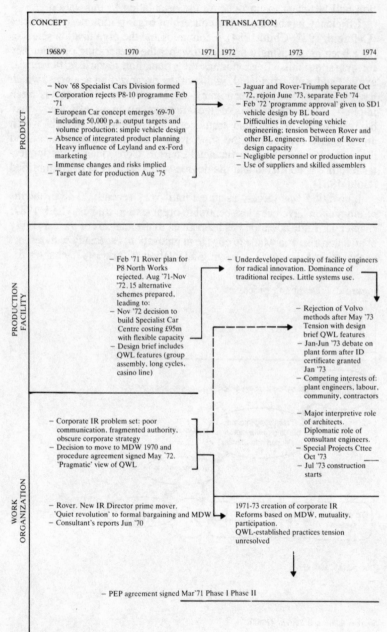

	CONCEPT			TRANSLATION		
	1968/9	1970	1971	1972	1973	1974

PRODUCT

CONCEPT:
- Nov '68 Specialist Cars Division formed
- Corporation rejects P8-10 programme Feb '71
- European Car concept emerges '69-70 including 50,000 p.a. output targets and volume production: simple vehicle design
- Absence of integrated product planning Heavy influence of Leyland and ex-Ford marketing
- Immense changes and risks implied
- Target date for production Aug '75

TRANSLATION:
- Jaguar and Rover-Triumph separate Oct '72, rejoin June '73, separate Feb '74
- Feb '72 'programme approval' given to SD1 vehicle design by BL board
- Difficulties in developing vehicle engineering: tension between Rover and other BL engineers. Dilution of Rover design capacity
- Negligible personnel or production input
- Use of suppliers and skilled assemblers

PRODUCTION FACILITY

CONCEPT:
- Feb '71 Rover plan for P8 North Works rejected. Aug '71-Nov '72. 15 alternative schemes prepared, leading to:
- Nov '72 decision to build Specialist Car Centre costing £95m with flexible capacity
- Design brief includes QWL features (group assembly, long cycles, casino line)

TRANSLATION:
- Underdeveloped capacity of facility engineers for radical innovation. Dominance of traditional recipes. Little systems use.
- Rejection of Volvo methods after May '73 Tension with design brief QWL features
- Jan-Jun '73 debate on plant form after ID certificate granted Jan '73
- Competing interests of: plant engineers, labour, community, contractors
- Major interpretive role of architects. Diplomatic role of consultant engineers.
- Special Projects Cttee Oct '73
- Jul '73 construction starts

WORK ORGANIZATION

- Corporate IR problem set: poor communication, fragmented authority, obscure corporate strategy
- Decision to move to MDW 1970 and procedure agreement signed May '72. 'Pragmatic' view of QWL

- Rover. New IR Director prime mover. 'Quiet revolution' to formal bargaining and MDW
- Consultant's reports Jun '70

1971-73 creation of corporate IR Reforms based on MDW, mutuality, participation. QWL-established practices tension unresolved

- PEP agreement signed Mar'71 Phase I Phase II

Source: Whipp and Clark (1986).

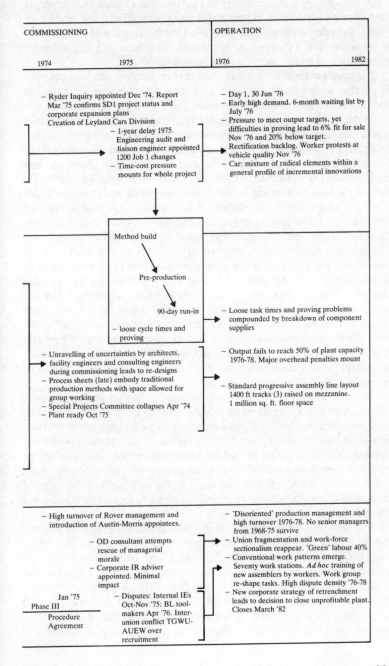

COMMISSIONING		OPERATION	
1974	1975	1976	1982

– Ryder Inquiry appointed Dec '74. Report Mar '75 confirms SD1 project status and corporate expansion plans
  Creation of Leyland Cars Division
  – 1-year delay 1975. Engineering audit and liaison engineer appointed 1200 Job 1 changes
  – Time-cost pressure mounts for whole project

– Day 1, 30 Jun '76
– Early high demand. 6-month waiting list by July '76
– Pressure to meet output targets, yet difficulties in proving lead to 6% fit for sale Nov '76 and 20% below target. Rectification backlog. Worker protests at vehicle quality Nov '76
– Car: mixture of radical elements within a general profile of incremental innovations

Method build

Pre-production

90-day run-in

– loose cycle times and proving

– Loose task times and proving problems compounded by breakdown of component supplies

– Unravelling of uncertainties by architects, facility engineers and consulting engineers during commissioning leads to re-designs
– Process sheets (late) embody traditional production methods with space allowed for group working
– Special Projects Committee collapses Apr '74
– Plant ready Oct '75

– Output fails to reach 50% of plant capacity 1976-78. Major overhead penalties mount

– Standard progressive assembly line layout 1400 ft tracks (3) raised on mezzanine. 1 million sq. ft. floor space

– High turnover of Rover management and introduction of Austin-Morris appointees.
  – OD consultant attempts rescue of managerial morale
  – Corporate IR adviser appointed. Minimal impact

Jan '75
Phase III
Procedure
Agreement

  – Disputes: Internal IEs Oct-Nov '75; BL tool-makers Apr '76. Inter-union conflict TGWU-AUEW over recruitment

– 'Disoriented' production management and high turnover 1976-78. No senior managers from 1968-75 survive
– Union fragmentation and work-force sectionalism reappear. 'Green' labour 40%
– Conventional work patterns emerge. Seventy work stations. *Ad hoc* training of new assemblers by workers. Work group re-shape tasks. High dispute density '76-78
– New corporate strategy of retrenchment leads to decision to close unprofitable plant. Closes March '82

Even the strategy literature fails to examine and explain how strategic directions are articulated and translated (see Quinn, Mintzberg, and James 1988). In order to unravel the innovation perspective it will be necessary to examine the expertise of the techno-structure and the roles of technocrats like engineers in mediating adaptation.

It is useful to define innovation-design as an iterative process occurring in parallel to ongoing processes at the shop-floor level as illustrated in the multistate model shown in Figure 2.2.

In the model there are four states (not stages!), and the fuzzy lines emphasize their convoluted, iterative, and often abortive processes (see Mintzberg *et al.* 1976). These four states are part of what Kantrow (1980) has termed the 'total process' of creating new products, processes, and forms of work organization. The figure depicts the articulation of strategic directions as commencing at the envisioning state and then gradually evolving through the translation state into the commissioning and operating of a new operating system. The classic research design of the 'before change' and 'after change' used in many studies of technical change only focuses on a very small fragment of the total picture. Moreover, the total process of strategic innovation can be examined as a design grid with the four states, and with the three key areas of the design of the product (or service), the production process, and the organization. The grid is shown in Figure 2.3 where details have been inserted from the innovation-design processes during the design of a new car, factory, and organization for a European speciality car (Whipp and Clark 1986). Although the details in Figure 2.3 are highly compacted, a careful examination will considerably add to an understanding of innovation-design as an orientation.

The orientation to innovation lays great stress on the longer-term future. Also increasing attention is being given to the presence or absence of a structural capability within an enterprise to engage in innovation-design (Clark 1987: chapter 5). The notion of structural capability is central. The implication is that we should give more attention to the puzzle-shaping and puzzle-solving capacities of the techno-structure (Nelson and Winter 1982) and to how the organization of internal expertise shapes the choice of directions in the large area of Design and Development.

The distinction between efficiency and innovation-design provides a useful means of highlighting the importance of innovation. The thinking behind the selection of this dimension may be illustrated from the seminal analysis by Abernathy (1978) of the dilemma facing American automobiles during the period from 1925 to 1970. Abernathy concludes that the two largest firms – General Motors and Ford – were committed to pursuing efficiency rather than innovation. They managed this dilemma by concentrating the public's attention on innovations of the outer skin of the car: its annual styling. Meanwhile the firms pursued long-term efficiency in the production of the mass of components and the sub-assemblies which were beneath the skin. Abernathy concludes that orientation towards efficiency was viable, whilst market and technological ferment were relatively slight. However, during the 1970s there were signs that the degree of required innovation was intensifying. All enterprises face a dilemma between orienting themselves towards being highly innovative or highly efficient (Lawerence and Dyer 1983). Abernathy demonstrated that US firms had resolved the dilemma by being oriented towards efficiency. Hence their vulnerability if the market conditions were to alter (as they did) towards a faster pace of innovation.

Frictionless structural transitions versus structural repertoires

In the vertical dimension the contrast is between theories and practices which depict transitions from one state to another as either:

(a)   the frictionless structural transformation of the existing structural features and capabilities into a different form as occurring relatively quickly and evenly;

or

(b)   the uneven restructuring and the possible failure of existing enterprises or their replacement by new entrants and/or enterprises formed through acquisitions.

This dimension is explicitly concerned with time-and-structuration. The contrast is between the time-free equilibrium models which dominate the organization sciences, and the use of event-based time-reckoning systems.

The four quadrants

Efficiency and frictionless transitions

The top left-hand quandrant (see Figure 2.1) is the orthodox mainstream which we labelled organization design theory. This quadrant refers to theories and models which emphasize efficiency in combination with frictionless structural transformations of firms. These theories are the dominant orthodoxies in economics and in the organization sciences. Until the 1950s the sole significant contribution came from equilibrium economics which were concerned with the macro level of the economy rather than with the internal operation of the firm (see Penrose 1959). In economic theories the frictionless transitions were achieved through the exits of existing firms and the entry of new firms, as well as through the 'black box' of transitions in the management institutions within existing firms. Competition between firms was theorized within an evolutionary framework, yet the Darwinian dynamics were rarely explicated. The analytic prestige of economics and its law-like theories shaped the subsequent development of the organization sciences in the late 1950s.

Differences between the economic theories and the organization theories emerged in the late 1950s with the work of March and Simon (1958) and the Carnegie School. Their theorizing was full of practical implication for events within the 'black box' of the firm and gave an impetus to the vision of a new discipline: organization design. Their analysis concentrated upon the examination of those structural features within the firm which facilitated or inhibited efficiency. March and Simon began as a socio-psychological critique of the rational assumptions in economic theory, and then introduced the notion of subjective rationality which located the individual within a structural environment whereby the objectives of the firm were shown to exert a powerful influence on individual cognitions and behaviour. The Carnegie School also rejected the analogy of the mechanical machine which was attributed to F.W. Taylor. Yet, ironically, they introduced a new machine analogy: the computer, its programs and software. There was a very close harmony between the vision of designing programs and the vision of designing organizations. Consequently their notion of structure as a repertoire of performance programmes was analogous to the directory of files in a computer. Within that important limitation the Carnegie School highlighted the significance of the cognitions of top-level decision-makers, and subsequently explored the role of cognitions in the political process of accommodation within enterprises. A significant outcome of their work was the notion of organization design. That is the deliberate choice of certain structures by powerful decision-makers based on their expectations of existing and future levels of uncertainty.

The organization theorists focused upon the elaboration of concepts and a prescriptive theory for adjusting the internal features of all kinds of enterprises. In principle their theories considered both short-run efficiency and longer-run innovation, but the concentration tended to be upon the short-run efficiency. Many organization theorists considered that the problems of innovation had been adequately covered in the theorizing about innovation of Burns and Stalker (1961) and its application to case studies in the British science-based industries (for example, rayon, electronics). The prescriptions of Burns and Stalker for innovation have been taken as the most widely-cited version of best practice. The Burns–Stalker framework implicitly underpins a great deal of current work: the ambidextrous organization, planned change (Zaltman and Duncan 1977; Kanter, 1984). However, a very significant part of the analysis by Burns and Stalker was neglected – their explanation of why transitions could not be frictionless! We shall return to this later in Part IV.

Organization design theory exemplifies the focus on shop-floor efficiency and the assumption that the design of organization and technology occurs through frictionless movement of variables. The auditing of an existing situation and its diagnostic interpretation is undertaken with a small number of variables whose values can be precisely interpreted through a law-like knowledge. The prescriptive theories define what should be done to achieve efficiency. There has been very slight attention to explaining why prescriptions are not followed, possibly because the underlying model of corporate survival is that of a revolutionary and Darwinian competition between firms to acquire and hold resources.

The many variants of organization design theory were the focus of various attempts at encyclopaedic reviews and of synthesis from which Galbraith's (1977) statement is seminal for its analytic rigour, clarity, and cogency. Moreover, Galbraith succeeded better than most alternative texts in conveying the image of designing as a process of envisioning alternative forms of organization to improve existing situations. Galbraith implied that designing was undertaken by experts in the techno-structure, and that the processes of commissioning and operating the new designs was unproblematic. Galbraith's perspective reinforced the possibility for the conscious design of organizations and therefore for shifting the role of the organizational adviser from a servant to a savant. However, the political processes of design and the political accommodation of the designing roles were unexamined (see Pettigrew 1973; Miles 1980: 176–80). So also were the problems of implementing any chosen designs (Clark 1975).

Efficiency and structural repertoires

There was a growing critique of key problems in the orthodox notion of organization design, particularly of the assumption of frictionless change. Certain developments occurred each of which represented an attempt to examine structuring as a process of a lumpy momentum in the broad perspective originally proposed by Crozier (1964). These developments tended, however, to be implicitly oriented towards efficiency, yet to gradually prise open the problem of innovation. That is why the arrow in Figure 2.1 penetrates the lower, right-hand quadrant. As indicated in Figure 2.1, these notions tended to be applied within the framework of efficiency. However, their implications are not restrictive and can readily be incorporated into an innovation perspective.

The three developments which we have selected were: the introduction of the notion of organizational configuration by Mintzberg (see Mintzberg 1983); the distinction between momentum and transitions developed by Miller and Friesen (1984); the concept of the structural repertoire (Gearing 1958; Clark 1976; Clark and Starkey 1988).

First, Mintzberg (1983) presented several rationales for adopting a perspective on structural archetypes within which it is assumed that the relationships between variables remain relatively enduring for longish periods of time. Five examples of the most widely-found archetypes were constructed as ideal types. Our purpose here is not to present their detail because the relevance of that is marginal to the innovation perspective, but rather to indicate how the rationales for an archetypal approach challenge the assumption of frictionless adaptation.

The theoretical rationale for the archetype perspective derived from the sociological studies of structural stability, especially the analysis of French bureaucracies (Crozier 1964, 1973). That rationale was apparently confirmed in observation and discussion with senior executives which indicated that they tended to shape organizational structuring by attempting to impose well-articulated templates rather than to consider the complex models of the organization design theorists. That particular rationale presumes that top leaders are able to impose structural templates and to recognize the structures of their own enterprises. Once chosen, structural forms tend to continue in their particular form and then to encounter the longer selection criteria which eliminates the forms with the lowest capacities to renew core resources. Archetypes also have a strong appeal to students because they embody a great deal of theory in a form which fits well with the learning and evaluation systems of the North American business schools. The archetypes are static, yet that feature has the virtue of emphasizing that the notion of redesigning structures has to be undertaken against the inertial tendencies of existing structural archetypes.

Second, in an important extension of the archetype perspective, Miller and Friesen (1984) both provided an empirical revision to the original five types and also investigated a large sample ($n=136$) of change episodes along three sets of thirty dimensions covering the environment, the corporate strategy-making, and the organization capabilities. Each dimension was scaled on a five-point scale of intensity and applied to the situation at the start and the close of the change episode to measure the differences. That process revealed a significant distinction. Almost three-quarters of the change episodes revealed that the established tendencies (for example, towards mechanistic organization) were reinforced, whilst one quarter was of episodes in which the directions were reversed. The former was labelled momentum and the latter was labelled transitions. The extent of momentum in the sample confirms the notion that organization structure cannot be satisfactorily decomposed into frictionlessly related variables. That interpretation was reinforced by a closer analysis of transitions.These were exceptional, wrenching episodes of some length.

Third, structural repertoires as a perspective invokes the analogy from the theatre and from American football of 'plays' which can be activated in anticipation of future requirements. Here the metaphor of plays can be contrasted with the orthodox mainstream which has relied upon the metaphor of the computer program. The repertoires of plays by theatre groups and footballers are not reproduced in a simplistic manner. Their reproduction is both precarious and has a tendency towards what Giddens (1985) has termed chronic recursiveness. Plays are open to considerable reinterpretation and often present opportunities for elaborations (see chapter 9).

Applying the notion of structural repertoires to organizations is highly significant because there will be sharp variations in the repertoire by level of the organization. Moreover, the notion of structural repertoire cuts across the widely accepted threefold division of marketing, production, and development in a manner which recomposes their relationships (see chapter 10).

At the strategic levels the time frame is likely to be both extensive and relatively uncertain; at the level of the techno-structure the time frame may be a little shorter and more focused; at the operating level the time frame could be quite short as in the case of the week for a supermarket outlet. Each of these levels will possess distinctive structural repertoires, yet their reproduction will be more awkward in the case of strategic innovation such as the launch of new product types. In long established enterprises there will be design cycles (Clark 1972a; Miles and Snow 1978; Clark 1987: chapter 5; Clark and Starkey 1988), but their unfolding might be quite variable (Whipp and Clark 1986). The unfolding and reproduction of the longer cycles will provide many opportunities for small variations, some of which might contribute to an overall reconfiguration of the structural repertoire to establish a new form. There is a strong probability that the regular, recursive unfolding of the structural repertoire provides for a degree of variation in the oscillations which enables collective learning to transform the original structure into an alternative, novel form. The emergence of new structural forms from within the existing structural repertoire might also provide an explanation for the improvements in productivity which are known to have accompanied certain situations in which the inputs of capital remained unchanged for several years, even for decades. That said, it may be suggested that the occurrence of dissipative transformations from the oscillations of existing structures in the repertoire probably requires a great deal of inventive agency from those who are orchestrating the structures performances. Clearly, that process of editing can be done from outside as in the case of American football where the coaches attempt to design and redesign the repertoire. Their agentic behaviours are probably paralleled by management services in those Japanese corporations like Toyota who have developed tightly-integrated structural repertoires.

The metaphor of a repertoire of structures is complex and the usage of the metaphor is only just commencing. It is therefore useful to take a simple and clear illustration of the structural repertoire, especially to take an example which illustrates how much more analytic potential lies in the notion of structural repertoire than in the current notion of a configuration. The study by Gearing (1958) of the structural repertoire of the Cherokee Indians is shown in Figure 2.4 as a stylized fact. There are four structural poses (Gearing's concept) each of which is known in a highly commonsense manner. Each pose implies sets of orderly relationships which are arranged by gender and by generation. Each pose is organized around a different element in the Cherokee social structure: the hunting party, the household, the clan segment, the tribe as a whole. Moreover, as can be noted from the linking lines in Figure 2.4 the members of a village move from one pose to another during the year. Consequently, there are likely to be occasional sharp transitions in prescribed role behaviour. For example, it may be necessary for young males to shift from passive behaviours to violent, exciting behaviours within a short period of time in order to adjust to a threat towards the village. These variations in the role behaviours of the same individuals are prescribed, yet require accomplishment. The example of the repertoire of structural poses can be applied to organizations (Clark 1975, 1976, 1985; Clark and Starkey 1988).

This quadrant represents an important development from the narrow structures approach of orthodox organization design theory. The archetypes (of Mintzberg, Miller, and Friesen) illustrates the limits of a narrow structure approach, yet is relatively static and lacks the potential for further development. The most promising line of analytic development arises from the theory of structural repertoires sketched by Gearing. When developing that line of theorizing it is necessary to distinguish the poses relating to: (a) the regular activities at the level of operating units in a framework of efficiency, from (b) the poses used by the innovation-oriented techno-structure in editing and maintaining the performance of the operating units (Clark 1976).

It is also necessary to examine whether the structural repertoire includes a capacity to couple innovation with design (Clark and Starkey 1988: chapter 4). Those developments are best undertaken in the quadrant dealing with structural repertoires jointly to innovation as a perspective.

Innovation and frictionless transitions

The top right-hand quadrant in Figure 2.1 contains studies and theorizing which are oriented towards the examination of innovation within a

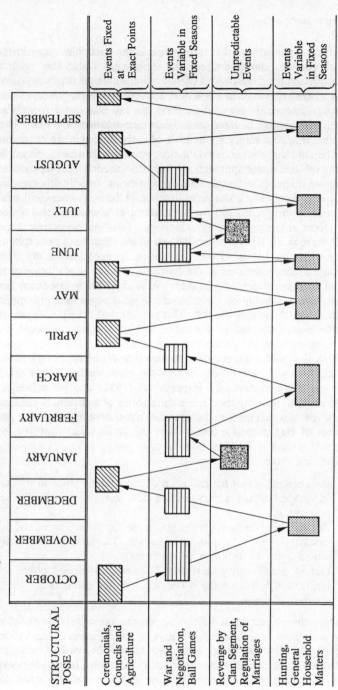

Figure 2.4 Structural repertoire of the Cherokee Indians

Source: Gearing (1958).

longitudinal perspective that tends towards the frictionless transformation of contexts and structures. The approaches within this quadrant frequently smooth the processes of innovation within the firm into stylized facts. There are two main kinds of contribution: law-like linear models and quasi-historical case studies. First, the law-like linear models are typically anchored in data sets. These are occasionally quite large samples, though a surprising number of early studies in the economics of technical change were based on more variables than cases. Examples of the law-like, linear approach are found in the early geographical and economic studies of technical change and in the use of orthodox organization design theory in a linear reconstruction of the evolvement of product and process design at Ford by Abernathy (see chapters 1 and 5). Second, there are a small number of individual quasi-historical case studies which have rarely been comparative and all too often invoke unexplained notions like 'real time'. However, the analysis of innovation is less developed than in the case of the economics of technical change or the spatial diffusion perspective. Also, it is difficult to ascertain how generalizations should be constructed and what ought to be the applied implications of the case studies. This section will briefly illustrate the combination of frictionless transitions with a focus on innovation with two samples from the generalizing approach.

First, this quadrant is occupied by the extensive literature on technical change, but will be typified through the early studies of the spatial diffusion of innovations by Hagerstrand (1952) and his subsequent exploration of how the time-space trajectories of individuals possess a recursiveness which tends to disrupt many novel developments. The main features of Hagerstrand's contribution can be captured from the four diagrams shown in Figures 2.5a, b, c, d.

The first three exhibits depict:

(a) the contention that the diffusion of innovations tends to follow a 'S'-shaped diffusion curve with a slow start, a take-off, and then plateauing;
(b) that diffusion tends to commence in metropolitan centres and then cascade downward to provincial centres. The same broad principle would apply to diffusion within firms;
(c) that in each locality there will be a neighbourhood effect with innovation following the connected communication nets over the shortest distances.

These three figures and their accompanying prescriptive implications have been part of the founding literature on the diffusion of innovations. Hagerstrand subsequently applied formal tests to the law-like theory and uncovered deviations from the theoretical expectations. Initially the deviations were explained by invoking the character of the innovation

Figure 2.5a 'S'-shaped curve

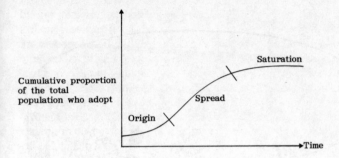

Figure 2.5b Hierarchy effect

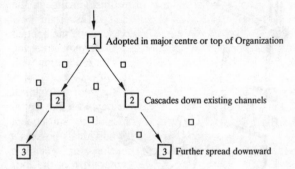

Figure 2.5c Neighbourhood effect

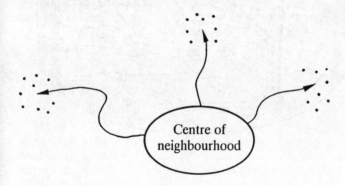

Source: Hagerstrand (1952).

Figure 2.5d Routinized everyday life

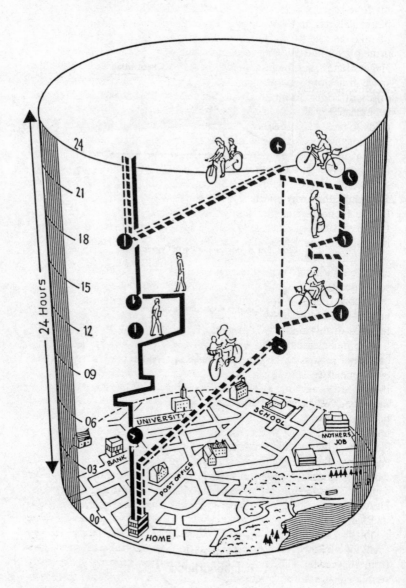

Source: Hagerstrand, in Chorley (1973).

networks. However, that direction was extended into an exploration of the time-space trajectories of individuals and this to the discovery that individuals tend to follow similar sequential patterns over regular periods like the day, the week, the shift, and so on. Figure 2.5d illustrates the principle for one day with respect to a husband, wife, and the one child. Hagerstrand's argument is that the aggregate of these individual time-space trajectories constitutes the terrain over which innovations are diffused. So, for example, a new form of urban transport system would be impacted by existing patterns. It is important to stress that Hagerstrand tends to understate the role of the intentionality of actors, and how that intentionality varies between individuals and within individuals. The principle of space-time trajectories is important, and will be carried forward for discussion in the next quadrant.

Second, one of the clearest examples of the linear models in this quadrant is provided by the Utterback–Abernathy framework (refer to Figure 1.1) and its application to the historical archives of Ford in the US up to 1975 by Abernathy (1978). This short, dense, monographic investigation combines the Utterback–Abernathy framework with organization theories (for example, Woodward 1965; Perrow 1967) to create a tight model which is applied in a reconstruction of events over seven decades. The explicitness of the guiding model and the rigour of the reconstruciton is seminal. However, we consider the study to be in need of substantial revision so this account concentrates upon the main features of the model by Abernathy.

The model used in the study of Ford assumed that in high-volume consumer sectors of the economy there are typical patterns of unilinear evolvement (see Figure 1.1) which car firms must adapt to and utilize if they are to survive. During the founding period of the automobile industry (1890s) the design hierarchy of the product was uncertain because the choice of a power source – the core decision – was initially uncertain. The early high uncertainty concerned the combination of elements which would constitute the 'horseless carriage'. The steam-engine was the dominant mode of power for American models until 1902 when the internal combustion engine emerged as the favoured power source. Thereafter, the elements of the car – the chassis, the brakes, the wheels, the body, and the power train were designed around the internal combustion engine and a design hierarchy of subsystems became established. During this early phase the suppliers assembled cars in very small lots. In Paris, for example, Louis Renault supplied customized cars from his corner garage. The key criterion for success as an early supplier was in terms of simple performance: how long did the car run before it required expert maintenance and repairs? However, with the stabilizing of a design hierarchy there was an 'inducement' (see Rosenberg 1982) to focus innovation on the blending of the components, on the reliability of performance, and then on the relative costs of buying and running.

Abernathy conceptualizes car firms as portfolios of productive units whereby a productive unit is a segment of the production process. A productive unit may be a component and/or a phase in the total production process. For example, an engine plant is often a separate productive units and so is the final assembly line. The notion is similar to that of unit operations in chemical engineering. Given the portfolio, Abernathy contends that units will all tend to shift from being arenas of novelty to becoming centres of efficiency. Efficiency as an orientation will drive innovation. This transition is irresistible, and those firms whose strategies do not fit the evolutionary pathway will disappear (see Whipp and Clark 1986: chapter 3).

The transition of the productive units can be traced along several dimensions as shown in Figure 2.6.

At the left-hand side the scale extends from highly fluid (that is, novelty) to highly specific (that is, routine). The headings across the figure refer to dimensions of the product, the process, and the organization. In the USA the speed of movement of productive units has varied. Engine plants moved earlier and more extensively because the final assembly line was the locus for the annual style changes, and because after 1926 – the end of the Ford T era – assembly lines tended to receive a variety of cars each with its own load of requirements.

The principles of the movement towards specificity reflect the theoretical influence of organization theory. The theory of social technology (for example, Perrow) is invoked to explain the steady movement of shop-floor problem-solving from many exceptional situations requiring skilled craftsmen to adopt a judgemental skill, towards a routine situation with few exceptions, each of which can be anticipated by the techno-structure which provides standardized rules for coping with the exception. Although Abernathy does not explain the role of the techno-structure, it may be derived from Perrow's theory. So it may be postulated that the techno-structure would itself start by having considerable power and would co-ordinate its activities through organic management systems. However, the theory indicates that the power would become constrained in established practices (which might be formalized), and that co-ordination would shift incrementally and cumulatively towards mechanistic, segmented forms of co-ordination. As a consequence the power of actors and subsystems to reconfigure the enterprise through structural intrapreneuring would be diminished. Therefore, as all productive units achieve specificity (in Figure 2.6) the techno-structure would set the rules for the operation of the shop-floor productive units. This format corresponds very closely to Mintzberg's (1983) archetype of the machine bureaucracy.

Figure 2.6 Portfolio of productive units

| | INNOVATION | END PRODUCT | INPUTS | FACILITIES/PROCESS | | | WORK ORGANIZATION AND SKILL |
				SIZE AND FLEXIBILITY OF FACILITIES	TECHNOLOGY		
1900/10 Founding era — Fluidity	Frequent and novel changes in the product	Ill-defined product. Cars produced for specific customers in the locality	Common type of components from regional suppliers	Very small in capital, space and employees. Small batches. Flexible units.	Little. Few general purpose machines for isolated tasks		Craft-based manual skills. Long cycle tasks. Piece-work
Specificity — 'Maturity' 1970s	Cost-stimulated incremental innovation. Very infrequent radical innovation	Functionally standardized components for the international market	Dedicated components. Integration backward to raw material suppliers	Large batches. Heavy capital involvement	Specialized, dedicated, and integrated technologies. Process and continuous flows		Monitoring tasks and process maintenance

Source: Based on Abernathy (1978), in Whipp and Clark (1986).

Although Abernathy's usage of Woodward's perspective on the role of market variability is very impressive, it does seem that the theory was used in a too restrictive manner (Clark and Starkey 1988). Reductions in consumer variability facilitated the usage of shop-floor technologies and of long-term planning methods in Ford. By the 1960s the customer choices (in North America) were increasingly channelled, choreographed, and orchestrated through advertising, through Hollywood, and the media. In a situation of oligopoly based on product differentiation the big three of General Motors, Ford, and Chrysler were able to move at a pace which fitted an orientation towards entrenching innovations introduced very gradually at the convenience of the manufacturers. It may be postulated that the market, which was highly homogeneous, moved more slowly than the modern Japanese market which is also homogeneous. Woodward's theory also postulates that the breakup of mass markets requires a shift away from the machine bureaucracy. The notion of the 'productivity dilemma' is used to signify that Abernathy observed that there were signs that the shift to specificity was being reversed. The theory of Woodward would require that the theorist would prescribe a reversal of momentum away from the 'structural rigidity' of a mechanistic mode to the 'structural flexibility' of the organic mode.

The postulates of the Abernathy model were expounded by powerful references to the biological metaphor of the life cycle from birth to maturity. Consequently, the policy implications of the model were that 'maturity' had been reached, and that the future would continue the past practices of competing on the basis of cost. That postulate was a comforting vision of the future for American firms, though it did imply that they should locate their most routine production of components in those parts of the Third World which offered the lowest wages and the highest tax concessions.

We have placed the interpretation and the investigation by Abernathy in this quadrant, because in our view there is too much optimism about the paces (in the plural) of structural transformation which are feasible from the archetype of the machine bureaucracy. The relearning requires significant, uneven transition, especially since the roots and routes for such a transition were introduced reactively (cf. Lawrence and Dyer 1983). Abernathy used inappropriate metaphors: the computer and the life cycle. By also invoking the metaphor of the computer and its programs, Abernathy assumed that enterprises could be structurally reprogrammed. A similar interpretation may underpin the more thorough going usage of the organic management prescription by Kanter (1983) in her fascinating analysis of ongoing changes in General Motors.

This quadrant is typified by optimism about the speed and ease of readaptive innovation. This predisposition is reflected in the attempts to revise and extend the orthodox organization design theory by simply adding a longitudinal dimension to an essentially static model. There are several examples, most of which rely on adding a calendar time dimension of orthodox contingency theorizing: dual structure theory and the notion of centres of innovation as 'reserves' (Galbraith 1977), and the adaptive cycle and strategic postures (Miles and Snow 1978). The extensive attempt to combine contingency theory with a longitudinal perspective is by Lawrence and Dyer (1983), who succeed in drawing attention to the significance of changes in the levels of information uncertainty, and in the availability of resources in the external environment as sources of problems of readaptation for enterprises, whilst confusing the means by which readaptation might occur. Their interpretation provides a neat bridge into the next quadrant. However, an important limitation of the interpretation of the empirical aspect of Lawrence and Dyer is that the distinctive features of the market structures within which North American firms are embedded remains outside the analysis (see Clark and Starkey 1988). So, the entraining potential of the slow-moving, homogenous US market cannot be compared with the faster pace of the equally homogeneous Japanese market. Moreover, the capability of foreign enterprises to penetrate the US market is not examined.

This quadrant represents an important development which has directed attention beyond the efficient operation of the shop-floor levels towards prising open the strategic decision-making processes. However, the reliance on applying orthodox organization design theory in an unilinear manner has been too restrictive and can be misleading.

Innovation and structural repertoires

Guiding principles

Developing this quadrant is the subject matter of the book. This section introduces the themes and provides an initial sketch of the main contributions so far. This quadrant treats innovation and efficiency as dilemmas which ought to be resolved through the activity of design. So the earlier orientation towards efficiency is replaced. So is the tendency towards a pro-innovation bias: the deliberate pursuit of innovation as an end. The dangers of that bias are evident in the much-vaunted British car: the Mini. The problem with the Mini was that it had to be sold to compete with the very basic, cheap cars produced in Britain by Ford. In the British contexts the Mini has been a financial disaster, and it is now known that Ford had recognized this defect in its rival more than twenty-five years ago! Pro-innovation is as dangerous as pro-efficiency. The objective of design is to resolve the dilemma between efficiency and innovation.

It would be foolish to claim that the new mainstream is as tightly reasoned as the orthodox organization design theory, because the new mainstream consists of a variety of perspectives all of whose development is still rather fluid. Their common feature is that they treat innovation in a temporally conscious manner which is beginning to provide a more systematic vision of how to develop a processual analysis which is sufficiently portable to be communicated and applied. The reader will have to use several perspectives in a trial-and-error form of learning. Moreover, there is an opportunity to develop a new form of methodology combining analysis and action (as Checkland 1981) which can replace the simplistic, singular notion of best practice as a clear template which ought to be imitated.

The stimulus for the attention now being given to this quadrant has already been mentioned in chapter 1. The analytic problems are to describe the new forms of market situation which seem to be replacing the large volume markets associated with 'Fordism', and to characterize the new forms of blending between organization and technology which can promote continuous, designed innovation. It is now becoming clear that the attempts of large private and public enterprises to adapt to the changed circumstances have already revealed considerably more puzzles and difficulties than the orthodox organization design theories had indicated. For example: the length of adaptation could be many years; the strategic directions might be the wrong ones; the time-scale of adaptation contains several different paces of transformation (Clark and Starkey 1988). Moreover, many highly-regarded examples of earlier innovation in organizations (for example, education and health) have been reevaluated to reveal that earlier innovative efforts had become encapsulated, or that the earlier claims were unreliable. The revelation of slow, uneven transition has not halted the plethora of best-selling studies of fast, almost frictionless changes. However, the prevalence of uneven transitions has provided an analytic opening for the population ecologists who argued that adaptation may occur more extensively through the changing composition of enterprises rather than through quantum leaps' by existing enterprises (see chapter 4). The population ecology perspective began to highlight the fixity of existing structuration and the role of blind chance (for example, random walks) in explaining corporate success. That conclusion had many implications which were uncomfortable for the dominant orthodoxy.

Overviewing the diverse collection of contributions to this quadrant is difficult and is bound to be more provocative than we intend. We have identified some guiding principles:

(a) there has been a shift of attention from the examination structure in its narrow sense to the examination of corporate knowledges and expertise defined very broadly to include the forms of knowledge, their vintage and relevance, the quality of the puzzle-solving regimes in the techno-structure and the overall capability of the organization at puzzle-solving;

(b) the crucial role of firm specific knowledge in innovation has been recognized and there are attempts to describe that knowledge in a systematic fashion. For example, by examining the languages used and the forms of cause-map (Weick 1979) which are implied. The issue of describing organizational recipes (for example, Hall 1984) and examining how the symbolic and cognitive structures are reformulated is now central;

(c) concern with knowledge and expertise has shifted attention from within the organization to examining the organization in its role-set of interacting organizations in order to discover poles and chains of innovative activity;

(d) the role of the market and of supply chains has become of increasing significance. The interfaces between suppliers and users has become a focal area of examination;

(e) serious attention is being paid to describing and examining the structural mechanisms through which inertia and momentum unfold and the means by which innovations might be inserted;

(f) explicating the temporal dimension as an array of complex routes and tracks for organizations and for innovations has become a major challenge to theory and to data-processing techniques. For example, the typical time-pattern for the diffusion of an innovation within a society might be the key clue to describing and explaining aggregate investment in innovations and the degree to which they are utilized.

These principles will be unravelled and developed in the book.

Five issues should be noted in this chapter:

(a) the interpenetration of levels of analysis and the transformation of the systemic to process models;

(b) the relevance of reconstructions of the past and of revisionist historicizing within organizations;

(c) the place of international comparisons of economic strength and the interest in the international transfer of innovations between leading economic states;

(d) the debate over a new techno-economic paradigm and the claim that new principles of international economic regulation are being evolved;

(e) the treatment of knowledges and cognitions as the building blocks of analysis coupled to the utilization of 'fuzzy knowledge' in prescription.

Each of these issues is significant to the guiding themes (above) and to the development of an innovation perspective. The issues are briefly explained below.

First, there is a shift in the level and scope of analysis from a tight focus on a narrow range of variables at the shop-floor levels to a broad, contextual 'mapping' which situates whole enterprises in their interfirm networks and gives attention to the international division of labour. The focus is upon the interpenetration of the macro global, the meso institutional and the micro organizational levels

Second, the issue of the historical dimension and its relevance has been raised. There are attempts to develop an approach which combines the reconstruction of the long-term past with revisionist corporate histories which deliberately challenge those sagas and myths which might impede transformations. The rise of Japanese economic power in the 1970s triggered much of this quasi-historical analysis of organizational capacities and has revived an interest in the business schools in macroscopic events like new techno-paradigms, long-wave patterns, depressions, and conjunctures. This issue is very evident in recent American attention to innovation. Thus, the self-confident narrative of the rise of the managerial strata in the USA as the visible hand of capitalism is placed in a rather different perspective by recent studies of the USA and its future. For example, Abernathy, Clark, K.B., and Kantrow (1983) challenged the continuing relevance of cherished managerial practices and severely criticized the existing masters in business administration. Likewise, Piore and Sabel (1984) sought to demonstrate that contemporary American management is unaware that they have been benefiting from a climacteric shift in the basis of international competition which occurred more than a century ago, but that a new climacteric shift is underway. Their view of the future emphasizes the decline of mass production and the greater salience of customized goods and services produced in communities of connected small factories and craftsmen reminiscent of nineteenth century Europe. Piore and Sabel give only passing attention to Japan which probably explains their romantic vision of the decline of large-scale production.

There has been an increasing frequency of reference to the importance of an historical perspective, particularly of the analytically-structured narrative associated with the Braudellian approach and with the work of Wallerstein. The latter provides the underpinning for a best-selling analysis of America's problems by Kennedy (1987) which compares the position of Philip II of Spain in the early seventeenth century to that of contemporary American presidents and their advisers. One of the key problems with the historical approach concerns the handling of the constraining influences of the past upon contemporary agenticness, and the part played by unintended structural outcomes and transitions. Too much of the recent usage of an historical perspective has failed to address this issue (see Whipp and Clark 1986: chapter 2). Therefore the general approach adopted by Dore (1973) in the investigation of Japanese transitions since the 1850s is of interest. Dore attempted to assess the relative amounts of continuity, of foreign importation, and of unintended novelty. The latter is especially important and connects well with the current interest in the notion of dissipative structures as unplanned structural outcomes. These may be very important in understanding the Japanese case where the US occupation after Hiroshima may well have made unintended impacts on Japanese society.

Third, the issue of international comparisons of economic performance have become of greater interest, especially the in-depth comparisons between the major economic powers and the institutional factors which explain their relative ranking. Maurice, Sellier, and Sylvestre (1986) brilliantly compared West Germany and France to reveal the contextual strengths of systems of occupational qualification and performance in West Germany. Now they are examining the role of French and Japanese engineers in cumulating and organizing knowledge and experience which facilitates innovation. This has led to a new emphasis upon the transfer of innovations between economic leaders which is displacing the previous emphasis upon technology transfer from the great economic powers to the Third World. Recent studies of the transatlantic diffusion of innovations are being complemented by transpacific studies. Amongst these, Cusumano's (1985) painstaking reconstruction of the different strategies for innovation appropriation pursued in Japan by Toyota (that is, reverse engineering) and Nissan (that is, licensing) is full of insights, and reveals the extent to which imitation is a great skill. It is now recognized that the Japanese case is of the appropriation and reinvention of western practices to create novel departures (Clark 1987: chapter 13). These studies of innovation transfer are providing a form of understanding about the contexts of innovation which were previously thought to be unimportant. Also, studies are being initiated on the role of multinational enterprises in innovation transfer both within the firm (see Clarke, I.M. 1985) and by the ripples of diffusion in localities through the neighbourhood effect (see Figure 2.5c). It is being recognized that transfer does and should involve the reinvention of the innovation (Rogers 1983).

Fourth, the notion of new techno-economic paradigm based on microelectronics is being invoked as the harbinger of systemofacture and its replacement of machineofacture as the cutting edge of future directions. The new information technologies are already altering the character of transactions within and between organizations as well as impacting the character of employment. This cluster of technologies is having major impacts in the highly-expensive areas of corporate co-ordination, where legions of white-collar workers, technicians, and managers handled vast amounts of information, and used their tacit knowledge to retain some degree of power over their situation. The initial wave of central computers introduced in the late 1950s created centralized data bases for areas like wages, and provided relatively inflexible systems of production control, but the new wave of small, powerful networks of equipment which can be operated on a stand-alone basis whilst being part of a large-scale system is having a massive impact. The new information technologies and their software have had an enormous impact in distribution, especially in retailing and in supermarkets. The latter are an exemplar of the logistical integralism which is linking distributors into tighter chains with their suppliers through flexible adaptation on a daily basis. The case of Benetton is a significant example of how the control over design and distribution (see Miles and Snow 1986) shapes the destinies of many small Italian firms (cf. Piore and Sabel 1984; cf. Sabel 1988). The same effects can be observed in West German manufacturing where the Bosch corporation utilizes many small-scale suppliers in Baden-Württemberg (cf. Piore and Sabel 1984). The new technologies have revolutionized the standardization of information and have provided a means through which a focal firm can specify contingently-explicit contracts, and also share information with its dependent suppliers so that the whole chain can economize on raw materials, on inventory, and on employing manpower when the demand has declined. We shall return to the significance of these 'managerial technologies' throughout the book.

Fifth, the approach to the theory of knowledge of this quadrant differs from that of orthodox organization design theory. Knowledge is increasingly treated as a fuzzy product, important aspects of which are deeply buried in tacit understandings. Moreover, the focus on systemic processes requires the construction of cause-effect maps by the analyst. The same maps are of relevance to those involved in strategic innovation because the cause-effect cognitions of all organizational members are significant in affecting long-term survival. The focus on knowledge reflects the shift from machinofacture to systemofacture. Hence increasing attention is being directed towards the mapping of the firm specific knowledges and to their assessment. This approach is, in many ways, a development of the human capital perspective and an attempt to unravel an equivalence between corporate investments in equipment and their implicit investments in forms of qualifications and structuring. There is a special attention to corporate languages and a detailed analysis of the constructs used (see Barley 1986; Boisard and Etablier 1987). Existing cause-effect maps are being recorded and analyzed and increasing attention is being given to ways in which members of enterprises learn from deliberate and accidental experiments. This development is highly consequential and could lead to significant revisions in thinking about innovation. For example, some years ago Swedish critics of the Kalmar plant of Volvo were pointing out that certain plants did not follow the same principles as Kalmar. However, this new approach to innovation and language cautions against expecting a sophisticated management (such as Volvo's) to be slavishly following a limited template (that is, Kalmar) when they might be developing a more complex language which discriminates between contingently different situations. If so, then enterprises will also have to develop do-it-yourself ways of altering their cause-effect maps (see Hall 1984). Hence, the new focus is for 'prescription via a detour of diagnosis' in which there is considerable usage of explanatory perspectives and insights. The approach relies on the skilful mapping of the external processes and their interpretation. Policy recommendations for strategic readaptation have to detail the time-scales for adaptation and to give recognition to the actions of competitors.

Societies, institutions, and organizations

The development of the new mainstream requires a balancing and interpenetration between the three levels of the societal position in the international division of labour, the meso level of institutions and sectors, the micro level of the organization. This means that organizational scientists have to extend their analysis from the micro level of the organization into the institutional and international levels, and to consider the impacts of long-term techno-economic paradigms. For economists it requires a more determined attempt to prise open the black box of technical change and possibly to discover that it is a Pandora's box inhabited by Sartrian engineers who are always behaving as though there were few structural constraints (Latour 1986). Each and all these groups will have to shift from the ritualistic critiques of orthodox strawmen into constructive theory building and research centred upon innovation and design.

At the macro level economic contributions have played a leading role in structuring thinking about invention and the diffusion of innovations. Economists use a threefold distinction between invention, commercial innovation, and the diffusion of innovations. The area of invention and the relative role of the pull of the price mechanism versus the push of scientific decision-making is a long-established area (see Dosi 1984) which will be treated as a context in this book. The commercial exploitation of inventions through prototypes (Leonard-Barton 1987) and through scaling-up for large-scale production (Sahal 1981) has given a heavy emphasis to research and development, whilst neglecting design and development. We shall concentrate more upon design and development. With respect to the diffusion of innovations (Stoneman 1976, 1983) we shall give particular attention to the interests of the organization perspective whilst aiming to make a selective usage of certain key issues. The macro level has been reconfigured by the debate around the long-wave theories of Kondratiev (for example, Mandel 1978; Mensch 1979) and of Schumpeter on the notion of creative destruction. Revisions to the Schumpeterian theories of the uneven evolvement of capitalism have been at the core of the approach of Freeman and colleagues at the Science Policy Research Unit. Moreover, they have stimulated analysis of the changing international division of labour and the restructuring of capitalism by sociologists and geographers.

At the meso level two areas have emerged, each requiring further development: the significance of institutional cores and the role of interfirm populations. First, the role of institutional cores in inhibiting or facilitating innovation and exnovation has always been important, but its significance is being reassessed. Recent studies have done much to unravel the role of the City in the British context and to explore Japanese institutions. An important issue concerns the degree to which the agentic, interventionist actions of the state apparatus can and are consequential for the transformation of existing institutions. For example, it is clear that in the early history of western capitalism, Britain was enabled to reap the benefits of guaranteeing international finance, and that currently the Japanese are enabled to benefit from the existing political economy of international trade. What is less clear is the potential for alteration of these institutional cores. Hence, the interpretation of the rescue of Chrysler as a virtually inevitable process actuated by the hidden rules of American banking and financial legislation such as chapter 11 is both elegant and daunting (see Reich and Donahue 1985). Likewise, the recent examinations of the institutional cores of various European states and the attribution to their values and networks of the explanation for relative industrial power omits the consideration of the practical implications (see Maurice, Sellier, and Sylvestre, 1986). Second, at the meso level attention is being directed at the various types of linkage between networks of firms. The level of analysis focuses on the changing populations of enterprises which constitute interfirm networks (see chapter 8). Earlier studies of interfirm networks were overly static and gave too little attention to the changing composition and the restructuration of network. That limitation is being corrected. The more complex perspectives have sought to apply concepts like value-added chain and filière in an effort to uncover the locations of power and dependence as well as to unravel the relations between suppliers, assemblers, distributors, and designers (Miles and Snow 1986). The notion of chains of dependency has been applied to Benetton, Marks and Spencer, McDonald's, and Toyota in a series of useful studies of the long-term capabilities for efficient innovation. Filière as a perspective has had a more chequered treatment.

Initially French studies sought to examine how far its electronics industry was a coherent network and to establish the consequences of losing nodes from the network through the penetration of foreign firms. However, much more application of the filière concept is required in order to specify its usefulness and limitations.

Micro level analysis of the firm attention is focused on the whole organization and on its context: the firm-in-context perspective (Clark 1987). At the micro level innovation was largely developed by economists in the study of technical change, yet much of that work treated the firm as an incidental location rather than as worthy of investigation. Little attention was given to corporate strategy and its directions. Too much salience was given to the roles of research and development, initially with studies of expenditure and then with attempts to develop a recipe knowledge about successful and unsuccessful portfolios of research and development projects as in Project Sappho. The findings from these studies (for example, Hindsight) often provided an illusion of 'objective useful knowledge', whilst their conceptual rigour and the operationalization of the concepts was quite primitive (Mowery and Rosenberg 1979). However attention has continued despite disclaimers (for example, Coombs, Saviotti, and Walsh 1987). Much of this work has lacked a credible perspective on corporate strategy-making and held to an unjustifiable assumption about uncertainty and the future. Too little attention was given to the design and development which accounted for four-fifths of the budget previously categorized as research and development. More recently it has been recognized that design and development became a Japanese strength. The emergence of that strength over more than two decades was neither tracked nor recognized until recently (Rosenberg and Steinmueller 1988).

We shall concentrate upon the micro/meso levels apart from a short examination of the international diffusion innovations in chapter 7.

Part II

Innovation

Chapter three

Dynamic configuration and contingent specificity

Introduction

The chapter commences with a short discussion of some of the problems which have arisen from the objectivist approach to innovation – now referred to as 'the trouble with technology'. The trouble is that technology is equated only with equipment and consequently the role of knowledges (the 'ology') is neglected. The basic problems of approach are itemized. The next section presents a solution to those problems in the form of a general framework: the innovation configuration framework. This framework is intended to provide a perspective which highlights both the evolving and uncertain features of innovation trajectories whilst also introducing the soft determinism which is central to this chapter. The application of an archaeological approach to the antecedents of contemporary innovations is also central to the approach in the chapter. Soft determinism has been one of the features of the approach by Gille (1978) to the examination of the interrelatedness and compatibility of complexes of innovations. Gille has constructed examples of a technical complex (for example, the blast furnace), and has constructed a simplified scheme of the systemic connections between western forms of technologies in the first half of the nineteenth century. The chapter concludes with an analytic sketch of network technologies, especially of the computer-assisted, interactive approaches to management co-ordination and planning. This final section examines a very contemporary problem – the differing approaches to co-ordination which are implied by American conceptions of computer-assisted management and by Japanese conceptions of just-in-time. The contingent specificity framework has the great advantage of countering the dangers of the pro-innovation bias whilst also directing attention to the problems of availability and fit between innovations and their users.

Objectification and the trouble with technology

Orthodox approaches to innovation treated the innovation as an object

51

like seeds or equipment: objectification. This has led to four guiding principles in theory building and empirical research each of which require revision:

1. The use of the 'before change' and 'after change' format as a standard research design to examine objectified innovations.
2. The search for a single scale to operationalize technology.
3. The conceptual split between administrative and technological innovations.
4. The attempt to identify which variables seem to be correlated with the adoption of objectified innovations.

These guiding principles share a common, unsatisfactory conception of innovations, and each represents a facet of the 'trouble with technology' (McDonald *et al*. 1983). In the orthodox approaches to technical change and to organization design reported in chapter 1 the preference for quantitative, cross-sectional, large-sample research designs tended to induce the usage of simple operational definitions of technology which focused narrowly on equipment and therefore neglected the 'ology' part of technology. Too much theorizing and research in the orthodox approach focused on a very narrow range of phenomena which were extracted from their contexts in a manner which is both unhelpful to understanding and probably dangerous to practice. Moreover, too much attention has been given to the shop-floor level. Insufficient attention has been given to the 'ology' in technology and to examining the making and unmaking of the 'knowledges' by management, by the techno-structure, and through strategic innovation (see Part IV). This section summarizes and explains the four problems itemized above and leads into the new frameworks which constitute the core of the chapter.

First, the problem of 'before change' and 'after change' studies of innovation. The usage of simplistic operational definitions by the orthodox approaches in the economics of technical change (for example, Mansfield), and in organization design, shifted theorizing into a narrow frame which neglected the knowledge aspect of technology. That narrow framework was convenient, but gradually induced a tendency to attribute too much influence to equipment and too little to the long-term strategic development of technology as a total process (Kantrow 1980).

The problem of treating equipment in isolation and of examining innovation within a very narrow time-slice can be illustrated by taking a contemporary situation in the supermarketing side of distribution – the introduction of electric point of sale systems at the checkout. To understand the significance of electronic point of sale it is important to recognize the lengthy heritage of manual stock control techniques and their initial development in the American distribution industry during the late nineteenth and in the very early twentieth century. These manual procedures were subsequently diffused to Europe in the 1920s. For example, during a visit from Britain to the USA in the mid-1920s, Simon Marks observed these techniques in operation. On returning to Britain he introduced them into Marks and Spencer's to revolutionize the stock-holding policy of the firm and to utterly transform its product range to a narrow, market-led focus. The knowledge underlying the manual procedure – itself an innovation – required the usage of simple equipment for calculating overall figures. During the next four decades the innovation evolved incrementally and created a configured network within each outlet of a retail chain. By the 1960s leading retailers already possessed a vision of how more of that innovation could be embodied in equipment. However, the equipment and the requirements for its usage (for example, barcoding) did not exist.

The principles of inventory management, and defining product ranges based on the analysis of daily sales, were rigorously applied to the operation of supermarkets after 1945. In practice the bulk purchasing policies and low inventory policies of the supermarket chains became the template for modern pull systems: 'just-in-time'. The pulling was initiated by the customers. The major supermarket chains competed on the basis of minute profits on a massive turnover. They progressively minimized their inventory and maximized the pull effect on their suppliers through exacting purchasing arrangements whereby purchasing from suppliers was tightly connected to the consumers purchasing behaviour. The introduction of main-frame computers in the 1960s facilitated the transfer of part of the manual process onto new equipment, and its recording on print-out which could be analyzed through statistical processes. However, a great deal of manual activity still remained, including the pricing of each individual product. The updating of inventory control and the resupplying of the outlets from central storage depots contained a number of weak linkages. Those weak linkages inspired the search for new methods. The possibilities of electronic points of sale were conceived in the 1960s and provided the trigger for the progressive application of barcoding systems on products. Later, in the 1980s, electronic point of sale was introduced.

It is evident from the above summary that any before change/after change which purported to examine the impacts of electronic point of sale would tend to omit consideration of the long-term evolvement of an array of procedures. The constitution of electronic point of sale required a massive degree of collective agentic action between the distributors and their suppliers of goods, as well as collaboration with the equipment suppliers. These long-term processes constitute strategic innovation (see Part IV) and cannot be properly understood solely within the before change/after change time-slice. Moreover, electronic point of sale became both an entrenching and an altering innovation (see

53

Part IV). The new equipment and its configuration facilitated entrenchments of existing directions as well as alterations in directions. The extent of the impact varies between users. In the retail firms which have already developed extensive procedures the introduction of electronic point of sale provides the regional centre of any large chain with a continuous real-time analysis of consumer behaviour as well as the performance of all groups of operators. That data can be used to plan the resupplying for the next day on a continuous basis. Also, at the outlet it becomes possible to adopt finely-tuned manpower policies whereby the hours of work are tightly allied to requirements.

The example of electronic point of sale illustrates the heritage of the configuration of knowledges within which the new equipment is located. Moreover, exploiting those potentials reflects the agentic capacities of the particular firm because there are important differences between firms in the speed and success at adoption. The example of electronic point of sale also illustrates the growing linkages through equipment of the shop-floor and the techno-structure of corporate analysts and long-term planners.

Second, there was an extensive search for a single scale of technology. The search commenced with the scales of Bright (1958) and the more elaborate-descriptive frameworks of the sociotechnical system (as Emery 1959), yet soon focused on a narrow definition of operations technology as proposed by the suite of scales in the Aston Programme (see Pugh and Hickson 1976). The reliability of the operations technology scale is uncertain (Clark 1987: 93–5). The use of the operations technology scale did prise open some of the claims made about the interconnections between technology and organizational structure. Later studies have confirmed that in practice there will be many different combinations of the same operations technology with forms of organizational process and structure (for example, Barley 1986). The search for a single scale of equipment enforced a conceptual and theoretical split between equipment and other dimensions with which it was configured.

Third, the objectivist conception of innovations also led to an analytic separation between administrative innovations and technological innovations. This was widely accepted even though the empirical content was very slight (Damanpour 1987). The separation may have possessed some merits when equipment was concentrated upon the shop-floor areas, but now that there are extensive usages of equipment in areas of co-ordination and of design the separation is confusing. For example, to define computer-aided production management without reference to its administrative facets would be unsatisfactory.

Fourth, many studies of innovation relied on the sales of equipment as the point of reference for diffusion studies, and sought to identify the relative influence of a variety of factors like size and perceived profitability on the decision to purchase. This format of research had the apparent virtue of focusing upon an unambiguous event: time of purchase/adoption. Given the commitment of (American) research to variance rather than process studies, this format became very widely utilized. Yet, even with this focus there was considerable variability in the findings (Mohr 1982). Moreover, it was quite inaccurate to rely upon sales of equipment as an indicator of technology usage (Brooks and Kelly 1986).

These four guiding principles drove empirical research and its inter-pretation for more than two decades, at the end of which it was recognized that a serious review was required. Gradually the reviews began to synthesize the various parallel stands of enquiry in economics, economic history, spatial geography, and in sociology. Also, there was a growing interest in the longitudinal dimension of innovation. The major area of development has centred upon knowledge rather than equipment. Thus whilst the early studies of technology as a residual variable concentrated upon hardware, the growing focus has been upon the social software (for example, McIntosh 1985). Knowledge in various forms has become the focus: learning by doing, learning by using, tacit knowledge, economics of information, and so on. It is these developments which have sharpened the limitations of the objectivist approach and have led to the search for an alternative.

Dynamic configuration

The development of the new mainstream requires a substantial revision of the orthodox treatment of technology in order to reveal the new guiding principles. Previously technology was equated with machines. Conse-quently, the knowledges which influenced the emergence and evolvement of equipment were neglected. Moreover, the forms of organizing and cultural systems which shaped the directions of their usage were also neglected. A new framework is required which reincorporates the 'ology' part of technology and which refers to both the abstract and the tacit knowledges. The new framework should address the dynamic unfolding of the relationship between disembodied and embodied knowledges. Our framework is set out schematically in Figure 3.1 and it may be observed that diverse knowledges are placed at the centre of the figure and that their embodiment is illustrated by reference to four facets: embodiment in equipment, in raw materials of any kind (for example, software and tomatoes), in the built environment (for example, 'intelligent office buildings'), and in standardized operating procedures. The figure, which should be used heuristically, treats innovations as a configuration which is dynamic and which may also be conceived of as a bundle of elements. Our intention is to highlight the configuration and then to address the conceptualization of the bundle of elements. The disembodied knowledges at the centre of the innovation configuration have four features which require immediate amplification: their plurality and diversity; their enormous growth; their codification; their embodiment.

Figure 3.1 Innovation configuration

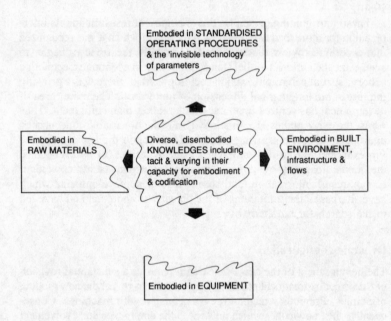

First, the plurality and the diversity of knowledges, and the relationships between that diversity within organizations, has been the centre piece for the notion of logics of action (see Karpik 1978). The concept of logics of action is an analytic construct for use by the observer to identify principles of action around which actors organize their attitudes and behaviour. So the logics of action provide linkages between collective and individual actions. The logics are often in the form of partial solutions to problems (see Nelson and Winter 1982). In any enterprise there will be a diverse plurality of logics and these may or may not be hierarchized and orchestrated in particular directions. The logics of action will, however, be specific to that enterprise – firm specific knowledges. Within any enterprise the quality of these knowledges will be a crucial feature of the human capital of the enterprise, and this cognitive dimension may be the most important competitive edge for some corporations

(Porter 1985). Also, within enterprises there are likely to be many problems associated with the revision and updating of these knowledges, particularly when new forms of embodiment arise.

Second, a salient feature of knowledge is its growth during the twentieth century. The significance of the incorporation of applied science into capital was indicated in the analysis of Marx and confirmed in recent studies of the role of institutionalized science within certain kinds of corporation (for example, Freeman 1974). The occupational mechanism through which this growth has been mediated and achieved has been indicated in the investigation of the growth of the industrial engineering and related professions in the USA and their role in the innovation-design processes (for example, Noble 1977). Growth has occurred across a wide variety of areas and more recently has concentrated upon the analysis of intracorporate processes through data-based comparisons and evaluations. There are many examples of this procedure ranging from education and hospital practices to construction. In the case of construction the vital tendering process around which future survival of firms is pivoted is anchored in a variety of data bases comparing alternative practices. These include the growth in knowledge about raw materials and their performance, especially cement.

Third, knowledges vary in the extent to which they are codified (Merton 1957). Initially attention was given to highly-codified knowledge like scientific laws as reported in journal articles and as described in patents. Pavitt (1987) suggests that technological knowledges are less codified than in science, though there may be clearly-articulated principles such as those used by system designers in the envisaging of new telephone exchange systems. Recent analyses of science and technology highlights the role and extensiveness of what has been labelled tacit knowledge. Knowledge is tacit when it has not been objectified into language, yet is essential for the skilled performance of tasks. Tacit knowledge is uncodified and therefore is transmitted through personal relationships and/or by close observation, for example, through the film analysis of activities. The codification of knowledge permits its transfer provided the standardization of the language is well developed, and provided that the potential users have been trained. There is nothing automatic about the linkages between codification and diffusion because diffusion has to be accomplished (cf. Boisot and Child 1987). Codification into knowledge systems is a major concern, especially of certain domains of expertise (Mackrimmon and Wagner 1986).

Fourth, knowledge can be embodied. That is to say, knowledge can be incorporated into artefacts. Marx's notion of machinofacture articulated his observation that the craft skills of employees could be transferred to self-regulating equipment which would then provide a more uniform standard over long periods. Likewise, the more recent discussion of systemofacture (for example, Kaplinsky 1984, 1986) refers to the transfer of white-collar cognitive skills to computer-based systems.

The embodiment of knowledge into capital goods has a long tradition in analysis, especially in economics. Therefore the recent development of theories which attempt to present an equivalence between the embodiment in capital goods and the embodiment in the creation of stabilized organizational practices is of particular interest. In Figure 3.1 there is an assumption that tacit and formal knowledge can be applied to the transformation of the four facets. It is also the case that the performance of the artefacts can facilitate concepts of knowledge (for example, the computer and artificial intelligence). The most frequently cited examples of embodiment refer to equipment, but embodiment is also applicable to raw materials, to standardized practices, and to spatialized constructions.

Figure 3.1 situates knowledges (in the plural) at the centre of the framework and shows an interactive relationship along four directions: to raw materials, to equipment, to buildings (and similar) and to organizational practices. The innovation configuration is a heuristic schema in which there is a dynamic relationship between disembodied and embodied knowledge and between each of the four facets. This section aims to highlight the significance of the previously neglected facets and to emphasize their relational qualities. Each of the four facets will be briefly examined.

First raw materials. It may be noted that raw materials have often been neglected, especially in the organization sciences. Apart from the analysis of the introduction of corfam into shoemaking (for example, Perrow, 1967), the stabilization of raw materials as diverse as cement and the tomato has been neglected. Likewise, the consequences of new raw materials have been dealt with in a very fragmentary manner which reflects the exaggerated attention to time-free theorizing and cross-sectional research designs. It is relevant to illustrate the importance of raw materials by referring to the massive developments from biotechnologies which relate to the production and distribution of food and drink. The enormous development of foodstuffs in the past two decades has been greatly enabled through the application of biological science. Modern supermarkets and their principal suppliers (for example, horticultural farmers, wine growers, chocolate makers) have become centrally involved in the 'design' of raw materials which increase the desirability of the products and the predictability of their operations (as Thompson 1967). Examples are to be found in all areas, yet are especially noticeable with fruit and vegetables. The modern tomato is a little noted, but important illustration. One consequence of the embodiment of scientific knowledge in the new raw materials has been the exnovation of existing craft concepts (for example, bakeries) of retail skill and their replacement by new skills based on combining the ingredients of raw materials (for example, types of dough) with new equipment (for example, baking ovens). Also large areas of knowledge about raw materials and their conceptual basis have been transferred and localized so that new knowledge is now conceived and authored by experts in the techno-structure and then diffused through various forms of corporate training (for example, use of video learning). At the shop-floor level of the supermarket the new raw materials permit the reconfiguration of shop-floor skills, and their homogenization to create increased flexibility for the management in the deployment of staff. The homogenizing of the differences in tacit knowledge between the various sections of the supermarket (for example, provisions compared to fruit) has dramatically altered managerial tasks and control. The same principles of analysis can be applied to the construction industry where there has been a massive development of modern forms of concrete with known rates of drying and with known rates of performance in the life of artefacts like roads and buildings. One of the most important new raw materials is software.

Second, knowledge can also be embodied in buildings, especially in their layouts and the allocations of spaces to particular interest groups. The ways which knowledge silently structures layouts at all levels from the household to the city remained outside conventional enquiry until the 'archaeological' style of analysis proposed by Foucault. Hospital design, for example, reflects the cumulated power of medical professionals. Now there are specialist groups of technocrats with particular claims about the relevance of their knowledges to the flows within society (for example, transport systems) and within firms (for example, industrial engineering).

Third, organization practices are increasingly the focus of codified knowledge, especially from the management sciences, and more recently from the organizational sciences. There already exist an array of templates of practices such as the choice between functional and product organization. However, as already indicated, a large element of existing organization practices are a reflection of situated practices (Giddens 1979) which are still beyond corporate regulation.

Fourth, the embodiment of knowledge in equipment is the most widely-recognized area. The problem with the examination of equipment as embodied knowledge is to construct a useful means for describing the essential features. One very obvious approach is through sketches and photographs. Today it is possible to film equipment and record its actions on video. There is a considerable problem of describing and recording the most relevant features of equipment for analysis. It may well be considered that it would be useful to return to an earlier practice of sketching equipment and showing its interrelationships. However, there is a requirement for more compact, generalizable frameworks. The threefold distinction between types of equipment is useful:

(a) equipment which transforms a raw material to a different state. The raw material can be of any kind, including data and symbols;
(b) equipment which transfers raw materials between workstations. These may or may not contain equipment. In manufacturing transfer equipment is being increasingly used to connect transforming equipment and this principle is spreading to the spheres of administration;
(c) control equipment which can be applied to either or both of transformation and transfer equipment.

These three types represent quite independent dimensions and can be used to describe developments in systems.

Figure 3.1 is intended to emphasize the interdependent configuration within which innovation is located. By situating knowledges at the centre of Figure 3.1 the intention is to highlight the previous neglect of the cognitive content of innovation processes, especially the role of the puzzle-posing regimes (Nelson and Winter 1982) and of problems chasing solutions (Cohen, March, and Olsen 1972). In addition it is vital to emphasize that the knowledge dimension contains a high degree of chance in the directions which are taken in problem-solving (David 1975). There is increasing evidence to suggest that once directions are chosen they tend to be reinforcing because of the many interdependencies which become aggregated around a given configuration (see David 1986).

Dynamic bundles of elements: reinvention

The new wave of innovation studies undertaken in the 1970s introduced two important principles. First, studies of the perceptions of adopters revealed that they often found it difficult to characterize an innovation and to anticipate how its introduction might impact existing methods of operating. Sometimes there was a great deal uncertainty (Gold 1981). Second, innovations were increasingly conceptualized as heterogeneous complexes rather than as homogeneous entities. This was explored by reconceptualizing innovations as bundles of elements. These two principles were combined in the issue of the multicharacteristics of innovations.

First, current studies treat the players perceptions as a crucial feature. The perceptions of the suppliers and users are essential starting points in any analysis, particularly their conception of the shape and uses of an innovation (Eveland, 1981). It might seem that the shape of an innovation is relatively uncomplicated. For instance, an American economic historian might walk through the collection of sawmilling equipment at the Smithsonian Museum in Washington and muse over the very extensive role of hard and soft woods in the early days of the settlement of North America (Pulos 1983). Also he/she would, no doubt, recall the seminal analysis of sawmilling equipment and ponder over the speed of cutting which created so much sawdust (see Rosenberg 1976). A British economic historian looking at the same equipment would probably recall the significance of the absence of wood in the British context and note how fast speeds often meant great wastage. However, most British visitors might be puzzled over the placing of sawmilling equipment in a museum at all. So there are at least three quite different perceptions, but would any of these correspond to what the users of those sawmills perceived?

The problematic nature of perceptions in the economic analysis of technical change has been examined by Gold (1981), yet the significance of this problem still requires emphasis because there is the basic problem of rationalizing perceptions after the event. For example, in the perspective of the late 1980s it is quite evident that there has been a massive growth in small, desk-top personal computers with immense capacities for storing, processing, and even transferring material. Yet, less than four decades ago a leading executive in IBM could imagine that the world would require a relatively small number of very large, mainframe machines. Also Rosenberg and Steinmueller (1988) have recently observed that American industrialists failed to recognize the extent to which the Japanese were innovating because the perceptual sets of Americans were oriented to low wages as the sole causal factor. Likewise Whipp and Clark (1986) report that the key set of British automobile experts who visited the Kalmar plant of Volvo in Sweden concentrated upon the routine mechanical features rather than the highly innovative electronics of the guided pallets. They also concentrated upon why such a system would not fit the labour relations contexts of Britain, but were unwilling to systematically examine those contexts and conceive their transformation. Consequently they failed to appreciate the thinking which lay behind those particular embodiments.

We should expect to find that the perceptions of what innovations are – their shape – and of their uses, vary widely between potential users and that these differ from what the academic in the role of a detached analyst might imagine. It follows from these points we should view retrospective histories which treat innovations as systems and as trajectories with care.

Second, the exploration of innovations as bundles of diverse elements was anticipated by Schumpeter, yet the early post-war studies emphasized the homogeneity of innovations. The new approach conceptualizes innovations as bundles of elements which have the following characteristics:

(a) they form a configuration with a core shape and uses which can be separated from secondary uses (Mohr 1982);

(b) the configuration has a shape and uses which will be perceived differently by various groups of users, even by different groups in the same firm;

(c) the configuration may seem to be tight, but can be unbundled when the users select only certain elements (Eveland 1981);

(d) the users may introduce new elements and so reinvent the innovation (Rogers 1983);

(e) the elements evolve dynamically and unevenly through calendar time in trajectories (Dosi 1984; Clark 1987).

Innovations are typically the combination and consolidation of many diverse elements.

The extent of modification and the evolvement of innovations is often understated, and the significance of this aspect is not fully captured by the important notions of learning by using, and learning by doing (Rosenberg 1976, 1988). One notion which has been advanced to highlight the modification of innovations is that of reinvention (Eveland 1981; Rogers 1983). The notion of reinvention has mainly been applied within the narrow frame of the adoption by single organizations of an externally-generated innovation. Although useful, that aspect understates the value of taking a more long-term perspective (see chapter 4). The processes of unbundling and reinvention are so important that some further elaboration is required to underpin the commonality of rebundling. Two examples will be utilized: the evolvement in the USA of Systematic Management (1880s) into Taylorism (1900s) and its transatlantic diffusion to Britain, France, and Japan; the evolvement of British rugby union and soccer into American football (1970s).

The various elements of Taylorism are depicted schematically in Figure 3.2 according to the tight bundling which Taylor envisaged. In the USA the unbundling of Taylorism commenced at an early stage. Nelson (1983) estimates that by the 1920s the new discipline of industrial engineering had recomposed the bundle to give particular attention to the systematic analysis of work flows, and to consider the workplace as a secondary level of investigation. Diffusion beyond the USA followed several different routes. In Britain there was extensive usage of methods of reward through payment systems which, until the post-war period, were often based on estimating rather than systematic work measurement. University-trained engineers rarely became involved with Taylorism. In the immediate post-war period the international chemical firm, ICI, was the centre for the constitution of an analytic approach to work design which became known as work study (Clark 1987: 282–6, 329–32), but that format was only partially diffused in Britain through the technician strata. In France there was much more attention to the performance of machines and layouts, possibly because the Grandes Ecoles

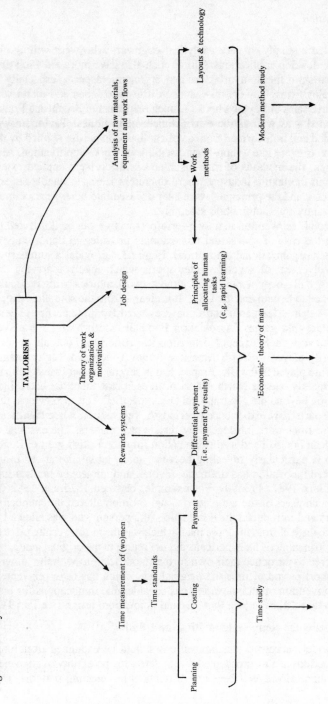

Figure 3.2 Taylorism: the bundle of elements

Source: Clark (1987).

provided a supply of very analytical engineers who were willing and able to develop applications. The French also gave more attention than the British to the human factors and ergonomic aspects, especially in the major firms. Taylorism was also diffused to Japan and there were extensive sales of Taylor's books – much more than in Britain and France combined – as well as the establishment of institutes. Taylorism was extended and transformed from a college-based expertise applied by the techno-structure into a shop-floor discipline of work simplification. Also in Japan, the methods of man-machine operation were applied extensively in the textile industry where operators learnt to handle several machines, and the principles were later diffused into other sectors, most importantly into automobile assembly.

Second, reinvention in a long-term perspective can be illustrated by taking the case of sports and this example provides an opportunity to spotlight organizational innovations. Figure 3.3 provides a summary of the evolvement of several winter sports which involve carrying and kicking balls over a period of almost two centuries with particular reference to Britain and the USA. It is clear – from the smoothed trajectories – that at least four sports have evolved from the earlier violent, irregular, folk games: Association Football (soccer) which is played almost everywhere except North America; American Football which is primarily played in North America; Rugby Union and Rugby League which are played in the UK, France, and in several former British colonial areas mainly outside North America. In each case there has been slight diffusion beyond the geographical areas specified. It is also evident that these sports have experienced periods of reinvention when significant modifications were introduced by certain new players. The example of American Football and its differentiation from the British sport of Rugby Union is particularly revealing because small, initial alterations in the imported innovation had dramatic, pivotal, and immediate implications (see Clark 1987: 170–90). For example, between 1876 and 1882 the American users of the template of Rugby Union altered the number of players and introduced the designation of a person who was allowed to remove the ball from the loose maul which was the historic centre piece of the British game. The Americans did not return to the original game, but preferred to introduce their own further modifications with the 'down'. In a short period of time the original innovation has been reinvented.

Reinvention of an innovation has considerable implications for both innovation-diffusion (see Part III) and innovation-design (see Part IV).

Antecedents, complementarities, and limits

It is widely accepted that innovations should be examined over long-term periods at a macroscopic level to detect the possibility of an overall patterning. However, there are difficulties in detecting patterns. One

Figure 3.3 Carrying and running sports: evolving innovations

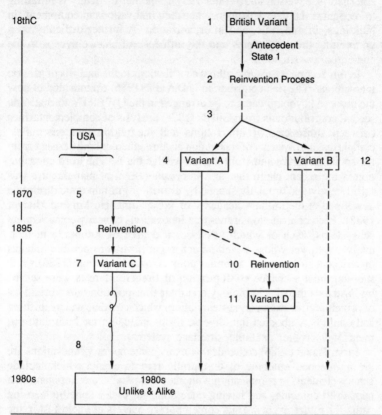

KEY:

1   British folk games
2   Appropriation of folk games by public school boys in England
3   Separation of the running-carrying game from the kicking game
4   <u>Formalization</u> of Rugby Union Football
5   Diffusion of Rugby Union Football & Association Football in the USA
6   American appropriation of Rugby Union Football & its modification
7   <u>Formalization</u> of American Football
8   Further evolvement of American Football
9   Schism of Rugby Union Football in England
10   Separation of professional rugby including modifications to Rugby Union Football
11   Formalization of Rugby League
12   Formalization of Association Football

Variant A = Rugby Union Football
Variant C = American Football
Variant B = Association Football
Variant D = Rugby League

Source: Clarke, P.A. (1987).

difficulty is that of imposing a coherence upon diverse innovations and failing to explain the selection environment from which the innovation emerged (see Nelson and Winter 1977). Another difficulty is in failing to recognize the interdependence between innovations in equipment, in buildings, in materials, and in organization. A further difficulty is in unravelling the antecedents and the influence of chance events on the contemporary situation.

In this section we examine three contributions to the analysis of generic innovations. The first contribution is David's (1986) examination of how the modern keyboard came to be arranged in the QWERTY format. The second contribution is from Gille's (1978) analysis of complementarities between subspecies of innovations and the notion that ensembles, complexes, and systems of innovation acquire inherent limits arising from the social structure and from the knowledge bases. The third contribution is the recent claim that the American system of manufacture was significantly and crucially shaped by a form of grammatonic discipline developed at the military academy of West Point (Hoskin and Macve, 1988). Each contribution argues that innovations typically contain many subsystems, each of which possesses a degree of autonomy in their evolvement, yet whose development might have considerable impacts on adjacent subsystems. A similar point is made by Latour (1988) when showing how a chance configuration of French interests were served by Pasteur's theories in biology and their transposition into a broad set of principles (for example, pasteurization) which could be widely diffused and could be embodied into diverse forms including the built environment of Paris (for example, drainage systems).

First, David (1986) contends that many contemporary innovations are the unintended outcome of temporally remote events which include chance elements in combination with multiple, parallel developments that happen to coincide, and having once done so, create specific learning paths for future events. Thus contemporary players are held fast in the grip of distant events and choices in which they were neither involved nor were their interests represented, yet those events circumscribe current decisions. David takes the example of the standard layout of type keyboards around the format of QWERTY and shows that other superior arrangements of the keyboard were available, but were not taken as the standard format. For example, systematic research by the US Navy in the 1940s demonstrated that the Dvorak Simplified Keyboard could be taught and its costs amortized in about ten days. So, how and why did QWERTY emerge as the standard solution? According to David, chance played a major role in the emergence of QWERTY as the standard arrangement. There were more than fifty patented attempts to create a commercial typewriter by 1867 and one of these involved Sholes and Densmire. Their version had its printing point located below the paper carriage, and therefore invisible to the operator who might not notice that the typebars had become stuck. It was awkward and time-consuming to unstick the typebars. Sholes arranged the keys in various ways to reduce the frequency of their compacting behind one another, and in 1873 the manufacturing rights were sold to Remington whose mechanics concluded (by chance) that the best arrangement was based on QWERTY. By 1878 Remington were selling their machines for about $125 and the entire sales for the USA were only 1,200 in 1881, and the vintage stock was only 5,000 units. Remington nearly went broke and certainly faced many alternative designs embodying different solutions to the keybar problem (for example, downstrokes, 1889–93) and hence different key displays. So, given the precarious hold of Remington on the diffusion pathway, why did QWERTY become the path dependent sequence? David emphasizes that all innovations face the environment of interrelated innovations and concludes that the users occupied a crucial role – by chance. The schools which were established for novice typists – initially men, but soon dominated by women – based themselves on the Remington machines. Likewise the early manuals. Consequently, the supply-line of trained operators were trained in QWERTY and this influenced the buyers and the newly-entering equipment suppliers. It was the buyers expectations about the equipment users which played a crucial role and which brought about *de facto* standardization by the 1890s. David emphasizes the importance of early 'windows' and their narrowness in shaping later pathways of evolvement, and the significance of complementarities between the subsystems around an innovation. This contribution draws attention to unintended creation of learning paths and therefore the crucial consequences of the choice of initial direction.

Second, Gille (1978) searches for macro level patterns in what he refers to as ensembles, complexes, and systems – each being a more macroscopic level of analysis. Gille contends that the embodied forms of innovations (for example, equipment) should always be examined in their socio-economic contexts. For example, the division of labour should be examined along with the legal system and the social structure. However, there is some tendency to overseparate the embodied forms in equipment from their social structural contexts and the economy. Gille is deeply concerned to explore the linkages between subsystems of ensembles and complexes as illustrated in the schematic analysis of the blast furnace shown in Figure 3.4 and the simplified scheme of the technical systems for the western world in the first half of the nineteenth century shown in Figure 3.5.

Each figure indicates the significant complementarities. This seems to be a rather systemic view of innovation, but Gille emphasizes that there are many subsystems, each of which evolves autonomously, yet might have far-reaching consequences for adjacent systems. The linkages

Figure 3.4 Blast furnace

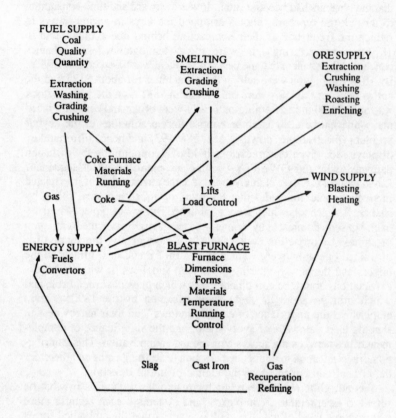

Source: Gille (1978).

between the subsystems will vary from tight to loose, though some may be more central, and over time the specific central subsystem will evolve. There is an evolvement of complex internal linkages so that an overall coherence is achieved. Coherence is a crucial concept which includes its associated division of labour and then combines these juridical and corresponding social institutions. These combinations vary in their openness to further evolvement. There will be a hidden hierarchy of diverse techniques and these will possess their own structural limits to further evolvement. Limits is a key concept. Structural limits might arise in one subsystem so that they constrain further development for several decades.

Figure 3.5 West European nineteenth-century technical systems

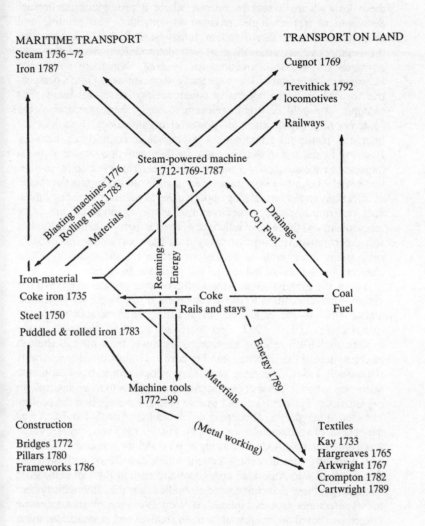

MARITIME TRANSPORT
Steam 1736–72
Iron 1787

TRANSPORT ON LAND

Cugnot 1769

Trevithick 1792
locomotives

Railways

Steam-powered machine
1712-1769-1787

Blasting machines 1776
Rolling mills 1783
Materials

Drainage
Coi Fuel

Iron-material
Coke iron 1735
Steel 1750
Puddled & rolled iron 1783

Reaming.
Energy

Coke

Rails and stays

Coal
Fuel

Energy 1789

Machine tools
1772–99

Materials

(Metal working)

Construction

Bridges 1772
Pillars 1780
Frameworks 1786

Textiles

Kay 1733
Hargreaves 1765
Arkwright 1767
Crompton 1782
Cartwright 1789

Source: Gille (1978).

For example, in the mid-1850s the rails in Britain had to be replaced so often that they became a major cost until the development of the steel rail in the Bessemer process. The larger significance of structural limits is indicated by the suggestion that in the medieval period Chinese technical knowledge was negatively impacted by structural limits, and these limits were absent in Europe where a prodigious intellectual development (for example, navigation) coincided with printing and promoted a coherent macro-system. It may be noted that Gille advocates the perspective on innovation of soft determinism. So, if the core attributes of generic innovations are ignored, then there will be an increasing disadvantage. There are many implications of Gille's perspective, especially in the notions of coherence and structural limits. For example, it would be highly relevant to apply these concepts to the contemporary development of powerful organizational networks for interfirm just-in-time approaches to logistical connections between Japanese firms, and to examine how far their adoption outside Japan is impacted by the absence of a similar coherence. That perspective implies that societal cognitive systems are very significant in shaping the choice of directions and in formulating opportunities to transform existing limits. Soft determinism as perspective also gives attention to the price mechanism, and the relative influence of costs is highly significant. Gille's soft determinism implies that many potential contributions to existing subsystems and overall complexes are produced, and their efficacy is assessed by the invisible hand of the price mechanism.

Third, the explanation of American economic growth after 1870 has given a central position to the development of the large, homogenized market, and to the emergence of a distinctive American system of manufactures (Clark 1987), yet there are puzzling elements in the explanation. Until recently the central issue was the dating of the full establishment of the interchangeability between standardized components (Hounshell 1984). The main contributions have come from economic historians who allocated organizational innovations to a residual role in explanations. Now Hoskin and Macve (1988) have applied the power-knowledge perspective (Foucault 1977) and have concluded that it was the genesis of accountability at West Point after 1817 which was pivotal. It is argued that methods of learning at West Point created and produced a meticulous grammatocentric system which its officers inculcated and then inserted into American armouries and certain railway companies from where these principles were embodied into the American system of manufactures as a discipline. At West Point the grammatocentric system was based on 'ubiquitous, written archives and examinations using mathematical grading' (Hoskin and Macve 1988: 38), and this developed a disciplinary power in its graduates which was diffused down a stream of key sectors and poles of learning. The origins of these disciplinary procedures of accountability in an élite institution with a special role in civilian life and their diffusion partially explains the development of cost and management accounting in the USA. Hoskin and Macve contend that this form of discipline was allied to the American system of manufacture and provided the knowledge base for its gradual development. We can interpret this claim as the insertion of a knowledge subsystem with distinctive thinking practices (Clark 1987: 221–6) adjacent to the attempts of the American government to break the structural limits (as Gille), and on the existing western systems of gun repair by introducing arms constructed from interchangeable parts. Existing subsystem of repair in Europe had reached the limits of the use of skilled armourers to repair defective arms. Consequently, easily interchanged components provided a remarkable advantage in warfare. In practice the standardization of gun making was a costly, very long-term investment which was promoted by the Federal government as an essential feature of the defence of the USA against European imperialism. The infusion of the highly-disciplined accountability gradually created a cadre of propagators who carried the notions of interchangeability into other key areas of the American economy.

Soft determinism and contingent specificity

This section presents the perspective of soft determinism and relates its basic principle to the notion of contingent specificity in the relationship between innovations and the organizations in which they are ingested. In organization studies there has been a long sterile debate over technological determinism. The notion of technological determinism can be applied to any analyst who claims that there *is* a simple causative and/or correlative association between the profile of equipment and the contextual and structural dimensions of organizations, and that connection is the direct outcome of the equipment or a change in equipment. That position is untenable as an observation about phenomena in organizations and in societies. There is no quasi-mechanical link between equipment and social behaviour (Clark 1972a: 149 f.) because as Touraine (1965) argues: orientations to work 'must be considered as autonomous elements subject to their own system of logic' and they cannot be considered to be a 'subjective reflection of the manner in which organizations function'. Yet a change does have a symbolic value when it is the outcome of a decision which alters the relationship of power or influence. Moreover, McDonald *et al.* (1983) points to the very diverse ways in which the same tool (for example, a farm tractor) has been used and embedded in different organizational arrangements. Invoking the technological determinism thesis tends to be connected with the ritualistic and vigorous defeat of the thesis.

Although invoking technological determinism is usually a strawman, the thesis does highlight the importance of separating two kinds of theorizing in organization studies. One kind of theorizing seeks to

71

describe and explain the existence of specific phenomena and to provide an understanding about their evolvement and impacts. Another kind of theorizing is the perspective of normative, prescriptive theories in organization studies which is concerned with what *ought* to occur to achieve a satisfactory level of economic performance. This position has been frequently referred to as the contingency theory, and it is the contingency theory which has been most frequently accused of harbouring tendencies towards technological determinism. Contingency theory attempts to specify the features which organisations should possess if they are to be efficient and to innovate. It must be emphasized that contingency theory is a normative, prescriptive theory of the means by which certain directions can be pursued. Contingency theory is therefore not a description or an explanation of how a specific set of decision-makers selected a particular direction and a means of attaining that direction. One early contribution to prescriptive, contingency theory was by Woodward (1958, 1965) who reasoned that the character of market demand – its degree of variability and uncertainty – was a critical variable around which the choice of other internal features (for example, planning systems, structures, equipment) should be arranged (see Clark 1987: 91–9). That perspective was not in itself committed to technological determinism: quite the contrary (see Galbraith, J.R. 1977). However, Woodward (1970) subsequently suggested that these are inherent requirements in the organization tasks which were associated with particular kinds of engineering arrangements, and it is that suggestion which does seem to suggest an inevitability. Some clarification about contingency theory and innovation is required.

Contingency theory, especially the perspective of Perrow (1967) has been applied to innovation in organizations (for example, Clark, 1972a, b, 1975) to argue that the *core attributes* of certain generic innovations should be adopted by organizations if they are to survive in the long run. This position is now widely accepted (Freeman 1974; Mohr 1982), yet needs to be applied with care. It is reasoned that the failure to adopt some of the core features of generic innovations will eventually incapacitate the rejectors. In the case of generic innovations like electricity, and its embodiment in portable equipment, the contingency perspective would anticipate that non-adopters can lose an economic advantage. Care is required because, as Porter (1985) demonstrates, organizations often acquire quite surprising bases for their competitive strengths. The core attributes of innovations are treated as the central features in the perspective of soft determinism. Gille (1978) contends that these core features must either be adopted or substitutes developed.

It is now relevant to examine the relational interaction between innovations and organizations. Earlier in this chapter innovations were presented as dynamic configurations in which the tool aspect and the use aspect evolve unevenly and sometimes in unexpected ways. This ongoing evolvement is likely to create periods of considerable ambiguity and uncertainty about the innovation itself concerning its potential, and also the capability of organizations to accommodate the innovation. At any one moment in time an innovation configuration will present itself to potential adopters as a complex bundle of elements whose relevance is *relational* to the pre-existing features of the host organization. The interactive encounter between the innovation configuration and the organization is both relational and contingent. That is, contingency is specific to the ways particular innovations are reconfigured to the specific requirements of the adopting organization.

The perspective of soft determinism and of contingent specificity is illustrated in the next section.

Network technologies: the case of computer-aided production management

This final section illustrates the use of concepts introduced in this chapter: innovation configurations which evolve unevenly; tight versus loose coupling of elements; uncertainty levels; and soft determinism. These concepts are applied to an exploration of one subset of the generic innovation of network technologies. The subset has been referred to by its promoters as computer-aided production management. The manifesto for computer-aided production management envisages that the central players in the problem of the co-ordination of activities within and between manufacturing firms should be production management.

In Britain there has been widespread concern about the rate and effectiveness with which the new equipment and software for information processing has been adopted and operated during the decade of the 1980s. One particular subset of network technologies may be taken to illustrate the themes of this chapter and the wider significance of the problems which are surfacing in the British context: computer-aided production management. This subset has been actively promoted by engineers, particularly by production engineers concerned with the control of manufacturing. One important definition of computer-aided production management describes it as a manufacturing technology that is primarily concerned with planning and controlling the material supplies and manufacturing operation of a business so that specified products may be produced by defined methods to meet agreed delivery dates so that people, plant, and working capital are used optimally (Cork 1985). Computer-aided production management is presented as a relatively tight package of elements which can be directed from a central position controlled by production. This vision of CAPM reveals an ordered complexity, yet it does not address the core themes of this chapter.

First, the evolving nature and contingent specificity of computer-aided production management is not sufficiently clear. In particular the problems and costs of collecting the basic data necessary to run a system of co-ordination and control which is centralized in the techno-structure is not addressed. That neglect reflects the influence of American predisposition to envisage software-driven equipment as the solution to all problems. In the USA that has led to the image of centralized information which is sufficiently updated and extensive to provide control in the operating units. The limits of that solution have been exposed by comparisons with the Japanese usage of software-based planning systems. Kochar (1988) suggests that the Japanese use software-based planning systems such as material requirements planning and manufacturing resource planning in a restricted way to plan the overall framework of production tactics whilst leaving the detailed shop-floor levels to a combination of just-in-time practices and work simplification (or Taylorism). Hence there is little need in the Japanese approach for detailed centralized information which requires considerable organizational discipline, is very expensive to update, and is deceptive as a vision of control. Kochar's analysis recognizes the combination of organization and of software-driven equipment in a contingency format. It follows that users of computer-aided production management should view both the various elements as being loosely coupled and as requiring embedding in an overall approach to information processing costs which combines organizational aspects with equipment. Our perspective suggests that computer-aided production management should be examined within the total innovation configuration which includes both:

(a) western variants with their historically-embedded predisposition to prefer software-driven equipment and centralized information control despite the costs of the decay of such information;
(b) Japanese variants in which the software-driven dimension provides the overall architecture for planning production, whilst devolving the adjustment to local contingencies to the subunit within a pull system.

This revised conception of the innovation configuration has dramatically different implications for the use of computer-aided production management. In practice the contemporary British scene can be best understood as a series of arenas in which there are ongoing struggles to blend American and Japanese systems of organizing with domestic predispositions. Second, the archaeological method of critique draws attention to the hidden antecedents and to the power implications they embody. The approach taken in this section is deliberately archaeological rather than genealogical because the aim is to prise open the hidden

74

assumptions and guiding interests which have created a specific path dependency (David 1986) and which may inhibit more fruitful solutions. Our argument is that the frequently cited and QWERTY-like genealogy of the sequence of production 'push' techniques which include program evaluation and review technique, material requirements planning and manufacturing resource planning should be the subject of a critique which exposes their silent origins about the best methods of organizing firms. If a genealogical approach is used, then modern techniques like computer-aided production management are presented as the unproblematic end points in a linear process: the pro-innovation bias. Yet, push planning as a system reflects the tendency in North America to locate the control of operations from the techno-structure. In the North American case the push systems are also sustained by the complementary interests of equipment suppliers, software houses, corporate leaders, and the professional groups associated with the introduction. For example, the American Production and Inventory Control Society was once the crusading agency for computer-based push systems. Until recently the American Production and Inventory Control Society had often been in the role of organizational gatekeepers enabling the rapid, relatively-uncontested introduction of push systems. Currently the American Production and Inventory Control Society is crusading with equal fervour for Japanese pull systems. The archaeological method of critique also reveals how far North American predispositions in push systems arrived unchallenged in the British contexts. In general the computer-based project planning techniques developed in large-scale American construction work were diffused into the leading firms of the British industry. Likewise in manufacturing the earlier manual planning by gross materials planning, which was a simple method of matching quantities of components to the end product was replaced by the computer-based material requirements planning. The manual system was associated with inventory planning and soon revealed the mismatches in existing systems leading to the introduction of material requirements planning. The material requirements planning systems in Britain were usually introduced into settings characterized by loose coupling and devolved decision-making (Clark 1987). Unlike the Japanese it was not possible to complement their periodic accuracy with shop-floor pull systems. Yet the British planners were drawn down the same trajectory of problem-solving as the Americans. Although material requirements planning balanced the inventory against future requirements, it also revealed problems of scheduling which in turn led to the development of 'closed loop material requirements planning' which was designed to facilitate comparisons between workloads and actual capacity. In due course the problems generated by that solution led the development beyond capacity planning in a major leap which incorporated the financial dimension into future plans through simulation: manufacturing resource planning. In Britain the application of these developments arose rather differently to the USA because the organizational equivalent to the American Production and Inventory Control Society occupied a different position in the organizational decision-making.

Third, computer-aided production management has been presented in terms of the focused interests of one professional grouping: those most concerned with the ownership of production problems. However, the general perspective of network technologies provides a better level of understanding, and identifies some of the key sources of uncertainty for the adopters – including production managers. Network technologies like earlier technologies have emerged sequentially through an evolutionary process in which the design and operation of subsystems are (were) adapted to the specific technical, politico-legal, and economic circumstances (David and Bunn 1987). Network technologies have not emerged fully blown, although some futuristic images have been invoked (for example, the paperless office). Moreover, the technologies have merged from a multiplicity of supplying sources. Their sequential and uneven development has and will raise many issues of compatibility between subsystems in the innovation configuration, and there may be gateway innovations (for example, open systems interconnections) which shape the evolvement of the network innovation. Network technologies are those forms of equipment, software, standard operating procedures and built environments (for example, smart buildings) designed to embody knowledges of co-ordination and strategic innovation-design in organizations (Kaplinsky 1984). These technologies convey coded meanings in the form of information based on predefined categories.

Network technologies contain various subsets. Three subsets will illustrate the range. One subset contains the media technologies (Rogers 1986) which provide connecting channels that are interactive, capable of being individualized, and are asynchronous in a manner which permits the storing of information in buffers which can be activated and responded to according to temporal convenience. The media technologies such as teleconferencing and local area networks depend on a critical mass of users and are strongly shaped by peer networks, because the quality of their usage becomes a crucial dimension. Media technologies are probably modified and adapted on a continuing basis to devise specific usages. Another subset is concerned with the storage of data in various forms. Initially these were widely applied to aspects of finance, especially the payrolls and similar, but later extended into subsystems such as the bill of materials. To some degree this subset came within the problem ownership of centralized data processing and accounting, though as their capability was extended (especially in the USA) they were increasingly viewed as the means of planning production and controlling the organization. Consequently, production controllers were active in giving salience

to their ownership of these problems. Moreover, certain software techniques, particularly those designed to remove production bottle-necks, fused
financial, sales, and production data into the same framework and thereby
confronted the competing domains of problem ownership between
finance, production, and marketing. Network technologies also provided
a capability for simulation and for design. Viewing computer-aided
production management as a subset of network technologies highlights
the significance of interaction, asynchronicity, and the horizontal flows
of disaggregated data to a multiplicity of subgroups, each with specialist
interests that should be networked.

Fourth, soft determinism and the contingency perspective assumes
that the generic innovation of network technologies have to be adopted
in some degree by all firms depending on their strategic objectives, their
manufacturing policies, and their product markets – a contingent
specificity. That is the perspective of soft determinism. The case illustration also assumes that this incoming generic innovation will significantly
reconfigure and alter the person-based systems and simple manual
systems of co-ordination which were already in position. The person-
based and manual systems were almost entirely operated by white-collar
workers, middle management, and by some members of the techno-
structure. Their demise was correctly anticipated in the seminal article
of thirty years ago by Leavitt and Whisler (1958). Child (1987) contrasts
the rising costs of the outgoing person-based systems of information
processing with the dramatically falling costs of equipment and software-
based information processing. Moreover, because of the high uncertainty
about network technologies as innovation, it is assumed that the extent
of the alteration of the pre-existing forms of co-ordination will be only
partially understood by potential users, and that at the level of specific
firms their enactment of the innovation will reflect particular politics
and power formations of problem ownership. It is assumed that the
organizational dimension of network technologies could be *understood
implicitly* and was therefore not problematic. Yet clearly, network
technologies are blendings of equipment with organization and raw
materials and are embodied in new forms of built environment. The
distinction between administrative and equipment-based innovations
simply collapses.

Chapter four

Stasis and dynamics

Introduction

This chapter deals with three issues which arise from the orientation to innovation presented in the previous chapter and concludes with a case study of the American fast-food sector to illustrate the themes of Part II and to provide a bridge into Part III.

First, we return to the distinctions between degrees of innovativeness as discussed in Chapter 1 and develop the fivefold typology to distinguish between innovations which tend to be:

(a) radical-altering in their requirements and in their impacts upon the user and on the suppliers

from;

(b) innovations which entrench ongoing tendencies and are therefore incremental

The distinction between altering and entrenching innovations is deceptively simple and highly consequential, not least because the analysis of the impacts of innovation has tended to overstate the frequency of radical-altering innovations.

Second, previous studies have identified ensembles of innovations which seem to possess an almost paradigmatic unity amongst the interdependent subsystems (Gille 1978; Perez 1983). It may be argued that these paradigmatic ensembles compete with one another, whereby one paradigmatic ensemble and its embodied knowledges is replaced by another ensemble. This phenomena may be referred to as substitution effect. The occurrence of substitution has been neatly exemplified in a comparison between two forms of sea-based transportation: the sailing-ship and the steamship (see Harley 1971). The substitution of sail by steam was much more lengthy than some accounts have supposed. It is therefore useful to take the 'sailing-ship' effect as a benchmark for comparing the key transitions and their durations against contemporary events – for example, in comparing manual and computer-based co-ordination systems of systemofacture (Kaplinsky 1984).

Systemofacture may be defined as the tendency to substitute equipment and its equivalence for cognitive skills. Whilst systemofacture is not the only formation taken by contemporary innovations, it is so highly pervasive that it requires particular attention in order to explore the difference between popular image and the contemporary diversity of situations. It has been argued that systemofacture is all pervasive (Kaplinsky 1984), yet the 'sailing-ship' points to the uneven entry of new complexes.

Third, the substitution theme raises the issue of the role played by innovations (radical and incremental) in explaining the shifts in the composition of populations of firms in the same line of business. The revival of population ecology models has drawn attention to the previous neglect of interest in the exits of established enterprises and the entrance of new enterprises. It is therefore appropriate to summarize our position with respect to the debate between the population ecology approach and the orthodox perspective in organization design of the theory of lowest information costs.

Finally, a case study of the slow emergence of McDonald's as the general template for fast food within the American fast-food sector anchors the guiding themes from Part II.

Radical-altering versus incremental-entrenching

This section examines the importance of the distinction between innovations in terms of their degree of radicalness.

The place of incremental innovation

An important requirement is to provide a framework which disaggregates innovations along the radical/incremental dimension and provides a means for replacing the orthodox tendency to imagine that all innovations are radical, massive disruptions. There has been a tendency to overstudy exciting, radical innovations (Rogers 1986). Our analysis suggests that many innovations which are essentially entrenching ongoing directions have been wrongly reported as radical. Moreover, the configurational approach suggests that many innovations may be complex blendings of facets in which the extent of radicalness may vary considerably. The degree of radicalness has to be assessed relative to the general state of the innovation and to the actual users.

We shall argue that one of the key dimensions to apply to innovation configurations is that which discriminates between:

(a) radical-altering innovations which reshape the entire configuration through the introduction of markedly different equipment, raw materials, forms of knowledge, and physical contexts. The consequence of such innovations is that existing competences become redundant and require exnovation. So, established directions are reversed;

(b) incremental-entrenching innovations which build on existing directions so that equipment is modified rather than replaced or knowledge is extended and reinforced.

The same innovation configuration may have an altering or an incremental consequence depending upon the purposes for which it is used and the state of the contexts into which it is inserted.

In the case of Computer Numerical Control (CNC) machines these can be used either as stand-alone equipment, or as integral parts of interconnected networks of equipment operated through computer-integrated manufacturing. In the former case the innovation may more often have an entrenching effect, whilst in the latter the tendency may be more towards altering the existing situation.

The radical/incremental dimension raises three issues. First, the Utterback and Abernathy thesis summarized in Figure 1.1 (see chapter 1), discussed in chapter 2 and examined more closely in chapter 5. How far is this thesis supported? We examine that issue and argue for the revision of the Utterback–Abernathy framework in chapter 5. Our argument is that the temporal distribution of bursts of radicalness is much more complex than the Utterback-Abernathy framework suggests. Second, the issue is whether the radical versus incremental distinction is more important than the distinction between administrative and technological innovations (Damanpour 1987). In organization studies the greatest attention has been given to the distinction between technological and administrative innovations. However, the administrative/technological split implies that innovations in equipment only apply to the shop-floor, core activities of the enterprise, and that administrative innovations only apply to the co-ordination of activities. There are two reasons for rejecting the usefulness of the distinction between administrative and technological innovations (cf. Damanpour 1987). First, as already indicated, the configurational approach has been constructed to reveal the blending of technology and administration. Second, because since the 1960s there has been an extensive diffusion and utilization of the generic technologies of information technology in the co-ordination of activities, in design and in integrating enterprises. Consequently an exponent of the technology/administrative distinction would have very great difficulty in deciding which category contained information technology. We contend that the technological/administrative distinction should be reconceptualized within the configurational dynamics presented in chapter 3. In our view the radical/incremental distinction is much more significant than the technology/administrative split because it is the blending of equipment within the innovation configuration which is important. Recognizing the blending is central to future practice and to understanding some of the differences between American and Japanese approaches (see chapter 7). There are growing reasons for believing that the blending is done differently in the USA and in Japan. The Japanese have focused on continuous, incremental modifications to their production systems, whilst American firms have concentrated too much on the introducing of radical, technology-based innovations (Hull and Azumi 1987; Brooks and Kelly 1986).

Third, in order to clarify the distinction between radical and incremental innovation it is useful to start by considering the problem of incremental innovations. The definition of incremental innovation is that the inputs to a system remain the same whilst the output is increased through alterations to the transformation system. One of the most important conceptions of incremental innovation derives from the seminal studies of 'learning by doing' in economic history (for example, Rosenberg 1976, 1982; David 1975) and the 'experience curves' reported by the Boston Consulting Group. The principles are similar. Researchers discovered that even when old equipment was used with slight modifications there were small annual improvements in performance which, when cumulated over a decade or more, could be very significant. The Boston Consulting Group concluded that each time an enterprise doubled its previous total output then there was a regular percentage gain in performance. By using the curve as an analytic device it becomes possible to imagine that firms could formulate corporate strategies through examining their relative position – past, current, and intended – on the learning curve for their product range.

Exponents of the ideas of learning by doing and of the learning curve tend to suggest that regular, incremental improvements in performance are part of the 'natural order' of events in enterprises and that incremental innovation is regular and unproblematic. Such a viewpoint is doubly flawed. On the one hand it fails to explicate the mechanisms and agencies through which small, continuous, and cumulative improvements are achieved. On the other hand it fails to explain the absence of incremental learning in many enterprises. It ought to be theorized that all innovation requires agency and the editing of existing concepts of knowledge about production techniques (see Giddens 1985; Clark 1987: chapter 7).

A case study of a British paper mill provides an interesting test of the problems of discriminating levels of innovation and also of the agenticness required to achieve small, continuous improvements (Scott-Kemmis 1983). The study examines a small mill within a large corporation whose performance had fallen behind the general level for the industry and had reached the point where its costs of production were exceeding marginal revenue. At that point the owners of the group introduced managerial changes which brought in individuals with a high degree of expertise about other similar, yet more profitable plants. The incomers aimed to improve performance, but had only very slight investment funds which they spent on purchasing control equipment to monitor the existing and ageing machinery. The new managers introduced a new cognitive set and applied its guiding principles to four facets:

(a) the price of raw material inputs was reduced without losing quality by changing suppliers. The new raw materials also reduced the energy bill;

(b) the oldest equipment – about one quarter – was scrapped. The remaining equipment was considerably modified, better maintained and control devices were added to monitor performance. The speeds of machining were increased substantially within six years and the frequency of breakages in paper-making was reduced;

(c) the flows of materials were improved by removing bottle-necks;

(d) large-scale changes in corporate organization and culture were introduced. These included setting higher standards, providing more training, making better use of information, briefing work teams, developing in-house scientific expertise which could be applied to the process and building trust.

These myriad, small-scale changes covered many interconnected facets and amounted to a systematic search by the management for the problems which arose from the new materials and the applying to these problems of expertise which facilitated trial and error learning. In due course existing cause-effect maps, at all levels, were edited.

The case study shows that there was a great deal of agentic intentionality. The only radical-altering innovation was in the introduction of control equipment onto the transformation and transfer equipment. However, the configurational perspective suggests that there were considerable alterations in the 'plant specific knowledge', and that these changes were used to blend a diverse array of facets. The new control equipment was an important ingredient in the transformation because its presence provided a tight cognitive framework of measuring and control loops. The alterations cannot be explained by simple reference to economic pressures (cf. Scott-Kemmis) because the case study points to the active role of corporate headquarters, to the customers, to the competitors and the suppliers. It may be concluded that similar changes in production technology might well occur in situations without dramatic alterations in equipment.

The case study probably represents circumstances which are found very widely. Incremental innovation is much more important than earlier studies had suggested.

Altering and entrenching innovations: Abernathy/Clark, K.B.

The distinction between altering and entrenching innovations has been persuasively examined by Abernathy and K.B. Clark (1985) who both unravel some important insights and also demonstrate the role of different kinds of innovation in competitive strategies (as Porter 1983a).

Abernathy and K.B. Clark propose a descriptive framework for categorizing innovations and for examining their varied role in competition between firms. The framework makes two basic distinctions. First, innovation configurations are not unitary phenomenon because innovations vary in the degree to which they disrupt, destroy, and render obsolete the already-established, firm specific, competences. Some innovations actually refine, enhance, and develop the existing competences. So, a distinction has to be drawn between altering and entrenching innovations. Second, the innovation can affect linkages along two core dimensions as shown in Figure 4.1. These two dimensions are on the production systems and their operation (the horizontal axis) and on the linkages between the firm and its consumers and the markets (the vertical axis). These dimensions are combined in Figure 4.1. The horizontal axis refers to the production system and the impacts of new innovation configurations vary from low/high and from entrenching to altering. With the vertical axis which taps the market linkages the impacts are also low/high and from entrenching to altering.

Figure 4.1 contains five examples of innovations introduced into the American automobile industry during the first third of this century. The Ford T is shown as disrupting existing linkages in the market and as rendering obsolete many existing modes of production soon after. In direct contrast the introduction of the electric starter and of the lacquer painting system are shown as innovations which entrenched existing market linkages and existing production systems. The remaining quadrants in Figure 4.1 are also important because of their consequences for competition between firms. The Model A Ford of 1927 is shown to have entrenched existing production competences yet partially disrupted the market linkage. However, there are two examples of innovations which entrench market linkages whilst altering production linkages: the introduction of the closed steel body and the introduction of the Ford V-8 engine. The comparison between altering and entrenching provides a marked contrast with approaches to innovation which stop at the moment of adoption or assume that the implementation and routinization must always be dramatic and altering.

The location of a cluster of innovations in the schema depicted in Figure 4.1 has a direct impact on the population of firms which compose the sector. The tendency for innovations to be entrenching will enable already pre-existing firms to maintain their competences and to benefit

Figure 4.1 Altering and entrenching innovations

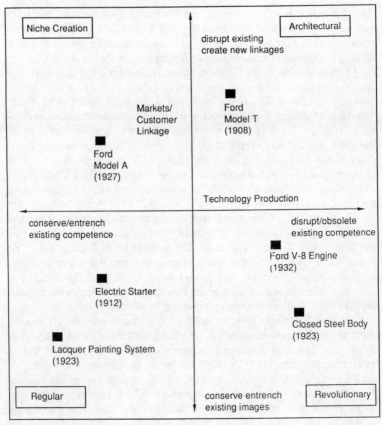

Source: Clark, K.B. (1985).

from cumulated, firm specific knowledges, whilst the periods when altering innovations occur provide opportunities for new entrants and renders obsolete the cumulated skills of existing firms. K.B. Clark (1985) demonstrates that in the American automobile industry the pattern of innovations in the 1920s, 1930s and 1940s directly influenced the opportunities which Chrysler had for challenging Ford's second position in the market. Chrysler were committed to innovative engineering, whilst Ford were committed to productivity (see Abernathy 1978). Consequently Chrysler narrowed the gap during periods of highish altering innovations, yet lost ground when eras of more entrenching innovation were predominant.

In any sector all four kinds of pattern are likely to occur at some stage and the temporal pattern of innovation is tightly linked to overall evolution of the industry. The objective of the distinction between entrenching and altering innovations is to highlight the possibility of alternate eras of altering innovations followed by unpredictable periods of entrenching innovations. The prevalence of either pattern for a period may influence firms to overentrain their competences to one situation rather than another, and to thereby inhibit their total adaptation. Clark and Windebank (1985) shows that in Renault the design groups which came to power with the radical innovation of the 4CV (1947c) retained considerable influence in later periods when the consumer preferences had changed. Similarly, Abernathy (1978) demonstrated that the corporate innovation policy of Ford had become one of slow, remorseless incremental innovation based on the progressive embodiment of skills and knowledge in equipment. Later analysis revealed that there had also been a tendency to develop protective systems to ensure the continuity of the assembly lines – the just-in-case tendency. Consequently Ford acquired a highly rigid organizational structure, and after the mid-1970s the existing skills and plants became obsolete.

The framework depicted in Figure 4.1 is intended as an analytic device for mapping the competitive consequences of different degrees of innovation.

Radical and entrenching: A and B types

Early thinking on the diffusion of innovations emphasized the general pattern of the 'S'-shaped diffusion curve, but subsequent research demonstrated that many innovations fail at an early stage and of those which do develop there are at least two major variants of the 'S'-curve (Davies 1979). The distinction is between the learning effects on the suppliers of entrenching innovations compared to altering innovations. The distinction is based on the supplying of process innovations. Innovations which are relatively simple, probably inexpensive, and can be produced in a turnkey format off the site are labelled Group *A* innovations, and their learning curve is displayed in Figure 4.2 as a dotted line. It may be noted that the learning curve rises quickly and then forms a plateau after which there are few reductions in cost. Group *B* innovations are shown in the solid line and require a longer learning period because they are quite elaborate and costly, and contain a high number of customized elements. For *A* type the learning effects as measured by declining labour inputs might be initially large, but these soon fall away. Whilst for *B* type there may be a very long learning period.

The research by Davies confirms the general importance of the distinction between altering and entrenching innovations.

Figure 4.2 Type *A* and type *B* innovations

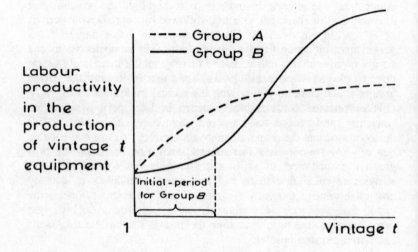

Source: Davies (1979).

Conclusion

This section has emphasized the crucial importance of the difference
between entrenching and altering innovations. In the relational perspective introduced in chapter 3 the meaning of radical and incremental is
relative to the supplier and to the user. The issue may be illustrated from
a contemporary situation in the service sector. In 1988 the largest retailer
in Britain – J.S. Sainsbury – announced that more than 80 per cent of
their goods were sold through electronic point of sale systems and that
they owned almost four-fifths of all electronic point of sale systems in
use at that time. The issue is: is the adoption and utilization of electronic
point of sale by Sainsbury's a case of an altering or an entrenching
innovation? The question is relevant because there has been a strong
opinion that the introduction of electronic point of sale into the Sainsbury
supermarkets was a highly-thought-out process of careful preparation
of staff (for example, video training well ahead of their introduction)
accompanied by careful scrutiny of the American experiences and of the
early British adoptions by Keymarkets at Spalding in 1979. If so, then
electronic point of sale may possess a strong entrenching impact. By
comparison there are other British supermarket firms whose movement
to electronic point of sale may well require such sharp changes in

managerial practices and in corporate ideology that its full operation might well have altering impacts.

Substitutability: the sailing-ship effect

Substitutability refers to the competition and symbiosis between two epochal innovations which serve similar functions such that one mode becomes less expensive than the other so that its usage causes the rival mode to be progressively displaced. The current illustration might be between manual and computer-based planning, design, and co-ordination systems, but the seminal illustration of the principle may be taken from Harley's (1971) reconstruction of the shift from freight transportation by sailing-ships to steamships over a four-decade period between 1850 and 1890. Harley reconstructs the length of the transition period and the probable role played by the substitution of coal for both capital and labour (and other factors).

Steamships became faster and more regular than sail-ships, yet the balance of advantage occurred slowly over four decades. The improved regularity of steamships provided a focusing device (Rosenberg 1976) for channelling the inflows and outflows of goods to plants and to markets. Speed and regularity also reduced the capital and labour costs per ton of goods carried for each mile travelled. So, although there was an increased cost of capital embodied in the cost of the steamship, and there was a need for a larger crew, these additional costs were more than offset by the increased speed of steamships. The continuing development of more powerful steam-engines permitted a lower coal consumption combined with increased speed. Coal consumption per unit of horsepower fell from 5 lb weight to 2 lb weight in four decades as the efficiency of the steam-engines was improved.

The economic basis of the replacement of sail by steam can be reconstructed by specifying the nature of the production functions and the substitutions. The production functions are illustrated in Figure 4.3a for a steamship carrying 2,000 tonnes of deadweight cargo in 1872 and in the accompanying tableau of Figure 4.3b. These reveal the trade-offs and the influence of the length of the voyage at three distances. The horizontal axis combines capital and labour on the basis of price as measured in pounds sterling. Because sailing-ships contain no coal, the production function for their services is shown at the calculated figure of £106, marked 'x' on the horizontal axis. The costs of sailing-ships are not influenced by the length of the voyage because there is no requirement to carry coal for their propulsion. A price-line for sailing-ships, which has been drawn through the 'x' point, shows the relative prices of capital, labour, and coal and provides a framework for judging the nature of the trade-offs and the influence of the length of voyage

Figure 4.3a Production function: sail and steam, 1872

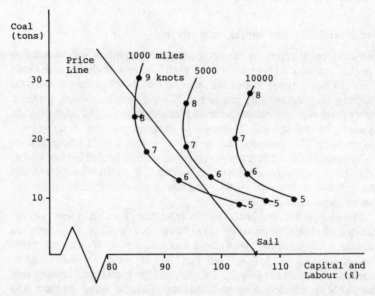

Source: Harley (1971).

Figure 4.3b Steam costs and length of journey, 1872

Voyage length (miles)	Optimal speed (knots)	Cost (£)
1000	7.0	100
3000	7.0	103
5000	6.5	111
7500	6.5	111
10000	6.0	115

Source: Harley (1971).

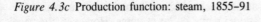

Figure 4.3c Production function: steam, 1855–91

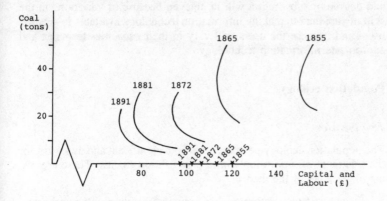

Source: Harley (1971).

on these inputs. To the left of the price-line steam was cheaper than sail, but to the right of the line steam was more expensive. The statistics indicate that sail-ships were cheaper for bulk cargoes on voyages of 5,000 miles in length in 1872. However, because of the increasing efficiency of the steam-engines in transforming coal the production function shifted to favour steamships over longer journeys as shown in Figure 4.3c.

These figures are based on the neo-classical approach to innovation. It is important to stress that these figures – assuming their general accuracy – were only partially known to the decision-makers of the nineteenth century. However, it is likely that a growing proportion of shippers would be aware of the general position and of the direction of costs. So, whilst some providers of the steamship service may have been premature in their entry as suppliers of the new service, the analysis by Harley cogently illustrates the consequences for all parties of the development of more efficient steamships.

The principle underlying the comparisons between sailing-ships and steamships – the sailing-ship effect – is of considerable relevance to the examination of contemporary innovations, especially those connected with information technologies: so called systemofacture. Information technology as a generic, contemporary innovation configuration covers many diverse elements in the same complex, yet overall it seems very evident that computers have massively reduced the costs of processing formalized information and increased the range of manipulations which can be used to approximate otherwise hidden processes. Therefore we can anticipate that information technology will continue to substitute for pre-existing manual, craft and white-collar systems. However, the pace and degree of substitution will be uneven because of variations in the skill of suppliers in making information technology available to specific users and because the users will vary in their capacities to ingest and appropriate information technology.

Population ecology

Two theories

The population ecology perspective is both significant and awkward for the treatment of innovation within organization studies. There are two major variants of the theory:

1. Strict ecological determinism of populations through natural selection on the basis of resource dependency and resource renewal. Therefore innovation arises from an external selection process which decides exits, survivals, and entrances amongst a population of organizations. The strict ecological determinism perspective.
2. Ecological pressures are pervasive but some corporations can consciously develop the capabilities for survival through the deliberate introduction of variations into the environment from the enterprise based on the intentions of vested interests like management inside the firm. The strategic innovation perspective.

Each variant emphasizes that managements have a free strategic choice of the directions which they may wish to take, but the strict ecological determinism perspective contends that the strategic choices are made in ignorance of the selection criteria which underpin success, survival, and failure. According to the strict ecological determinism perspective, it is environmental selection which shapes the fate of all kinds of innovation.

The population perspective makes explicit usage of the socio-evolutionary perspective which implicitly underpins orthodox organization design theory. The explicit usage of a natural selection model by the population ecology perspective postulates that some existing organizations will exit from the population and new organizations may enter. Also, the relative economic viability of existing organizations will be subject to irregular revisions in their ranking. More significantly, the strict version of the population ecology reasons that the innovations which organizations have to produce in order to survive are inputs into an environment which the 'visible hand' of management can only rarely anticipate.

Because the development of organization studies was an explicit attempt to prise open the economists treatment of the firm as a 'black box' (Penrose 1959; Rosenberg 1982) and to examine the machinations of the visible hand (March and Simon 1958; Chandler 1977) the problem of competition between organizations was largely ignored. So also was the problem of analyzing the varying balance between the size of the existing population of organizations and the availability of the resources on which they depend. Such problems did not seem to be very acute in the eras of economic growth from 1945 to 1970 and that might explain the low attention given to population ecology and to industrial organization. Organization studies implicitly espoused theories of managerial capability to discern the future and to adapt their organizations through deliberate planning and through managed innovation, therefore, the experience since the 1970s of a growing list of organizations (in all sectors) disappearing, being taken over, experiencing traumatic transitions, being new entrants, becoming sector leaders. These entrances and exits, combined with the massive programmes for internal transformation introduced by many large, established organizations have stimulated interest in the population ecology perspective and that perspective has been restated with an analytic vigour on the subject of innovation.

We shall examine the strict view in the next section before considering some of its limitations.

Strict ecological determinism

The strict interpretation provides an important counter to earlier theories which depicted corporate leaders as heroes capable of turning round almost any situation.

First, the notion of natural selection is central. In natural selection models it is postulated that organizations produce variants, that is, ensembles of their products (or services) including the image of the product, its price and design. These variants encounter pools of resources which may be very thin or may be very thick and bountiful. For example, Henry Ford's most successful variant was the Model T and that encountered resource pools which were bountiful and willing to provide resources of finance to Ford in advance of the receipt of the car. The relationship between the variants and the resource pool reflects hidden rules which have been labelled the selection criteria. These selection criteria are neither necessarily stable or easy to detect, yet reconstructing such rules lies at the base of the modern sciences. The selection criteria mediate between the type and number of variations to the type and size of resources.

Second, how adaptive are organizations? The population ecology perspective contends that the structures and problem-solving regimes of organizations always possess a very high degree of momentum in restricted directions and are frequently overcome by inertia. This feature has also been emphasized by the archetypal school (see chapter 2). According to the population ecology perspective, inertial structures are simultaneously a requirement for long-term survival and also an outcome of an externally-driven evolutionary selection process. Whilst the population ecology perspective does not deny the efforts within corporations by certain vested interests to achieve transitions in specific directions, it is emphasized that the capacity to adapt is greatly reduced when the pace of change in the selection criteria increases. The future form of the Darwinian selection rules is often unknown and, even when the rate of external shift in the rules is slow, the capacity of any corporation to adjust is always diverted by its internal politics of accommodation between vested interests. So the directions chosen by vested interests cannot be based on an adequate knowledge about the future form of the successful selection rules which transfer resources from the environment to the organization.

Third, there are many factors internal to the enterprise which generate and sustain structural inertia (see Burns and Stalker 1961: chapter 6). These include the dynamics of struggles between vested interests and the random matching of the strategic directions of the most influential élites to the requirements for obtaining resources from the external environment. The political relationships amongst vested interests constitute a local investment, and corporate ideologies often possess a quasi-hegemonic legitimacy which undermines counter moves. Also there are extensive sunk costs in both capital equipment and in standard operating procedures.

Fourth, the population ecology perspective challenges the widely-held view of rational adaptation theory which implies that variations in organizational strategies and structures are in anticipation of upcoming external threats and opportunities. This claim receives some support from Miller and Friesen (1982) who observe that fundamental transitions in organizations are rarely successful. Also, the exacting problems of transition were revealed in the indepth sample of enterprises from eight British sectors. The core issue here is the degree to which organization change is planned and can be controlled. Population ecology perspectives would explain the fact of changes in the rank order of particular sectoral populations, by reference to changes in selection criteria forcing exits and declines whilst offering new opportunities to a population of organizations which are inert and diverse.

Fifth, contrary to the assumption that innovations are produced because they are useful, the population ecologists argue that innovations are just produced and that innovation is a blind process. This strongly challenges the use of evolutionary perspectives by Nelson and Winter (1982) in the innovation and structural repertoire perspective where

cumulating experience to incorporate learning and to develop a repertoire are quite central. Organizational repertoires and skills tend to be
highly specific, and therefore vulnerable to shifts in the environment.
The population ecology perspective largely dismisses the relevance of
organizational approaches which seek to develop an expertise within
corporations to anticipate and act upon externally located changes in the
Darwinian selection criteria.

Sixth, population ecologists note that recent analytic studies have
shown that the rational pursuit of profit maximization is extremely rare
because, once founded, organizations elaborate many objectives.
Moreover, external criteria such as public accountability tends to
eliminate organizations which are low on accountability. This favours
organizations which achieve accountability through the institutionalization of reproducible structural repertoires. These repertoires may become
very extensive so that multiple routines can be accomplished. However,
because organizations become segmented hierarchically and horizontally,
there is a tendency for these routines to become fractured and so, during
the regular periods of the dormancy of some routines, they will decay.

These six features of the population ecology perspective provide an
important corrective to their neglect, but do they overpower the possibility
that there exist populations of organizations which have consciously
altered their repertoires in anticipation of changes in the rules through
which they acquire the basic resources of revenue and legitimacy on
which they depend? The crucial issue is the degree to which existing
enterprises proactively reconfigure their expertises and how they ingest
(or substitute) new capacities to obtain resources.

An important benefit of the population ecology perspective is that it
draws attention to long-term forms of economic regulation and to the
significance of their shifts. For example, there is widespread agreement
that the selection criteria for survival and for success have been changing
in the past two decades. Reference is increasingly made to changing rules
of international regulation for trade. Piore and Sabel (1984) contend that
between 1850 and 1870 there was a major transition from the rules of
mercantile trade at which Britain, as a merchant and naval power in a
useful geographical position, was enabled to excel, towards the strengths
of a land-based market homogenized by the agentic actions of railway
bureaucrats who simultaneously created regular transport and also fast
telegraph communications which 'wired' together producers and
distributers (Chandler 1977). It may be implied that American enterprises were constructed in an ecological space which favoured the
emergence of multiplant corporations pivotted around the graduate-trained
design expertise of the techno-structures which shaped consumer taste
to anticipate regular, yet slow innovation. Piore and Sabel maintain that
these selection criteria have been replaced by criteria which do not favour

the structural repertoires of most existing American enterprises and others which have sought to mimic their segmentation and just-in-case procedures. The survivors of the climacteric transition in the rules of economic regulation seem to include at least three diverse populations: the entrants from the Pacific Rim, the reconfigured western giants, and a host of new interdependent enterprises which organized finance capital and skills in innovation-design. The extent to which the strategic directions of any of these three populations is unintended and accidental is at the core of the current debate.

The value of the population ecology perspective has been in challenging the orthodox approach to organization design and in returning attention to the innovation.

Strategic innovation and the future

The issue is whether enterprises learn from their environments in an anticipatory, pro-active manner which leads them to change their repertoires in pace with their environments. Do the organizational élites engage in anticipating the future through strategic innovation? There are three elements in this issue:

1. The temporal patterning of changes in the key environments to which an enterprise has elected to become coupled or to which its enactment of the environment creates a socially constructed entrainment (Clark 1989).
2. The speed and fidelity of the learning mechanisms within the organization – the puzzle-solving regimes of Nelson and Winter (1982) – within the enterprise.
3. The responsiveness of the repertoire to designed, intended changes. How quickly can the repertoire be reconfigured?

These questions form the spring-board for developing a strategic innovation perspective (see Part IV).

The case for qualifying the strict population ecology perspective and for presenting a strategic innovation perspective can be derived from the explanation of the existing situation amongst organizations. There is widespread agreement that the selection criteria for success are shifting, and those shifts have crystallized the recognition of what were the previous criteria whilst leaving some doubt about the exact parameters of the new criteria. We are in period of transition. Contrary to the strict interpretation of population ecology there is widespread evidence of attempts by existing organizations to anticipate the future. The attempts are more evident in some countries and some corporations than in others; pro-activity varies.

The emphasis upon new criteria is more often mediated than direct, and the major form of mediation is through financial resource account-ability. In general terms this reflects the salience of finance capital and the activity of corporate acquisitions, resales, managerial buyouts, plant closures. There has been extensive ablation of plants, divisions, and employment in almost all major corporations and in many public enterprises (for example, health and welfare). The role of finance capital takes various forms and in some cases this is through the privatization of parts of public organizations and in managerial buyouts. These approaches represent a radical reposte to the organizational sciences (cf. Bennis, Benne, and Chin 1961). Education has become a major area of activity with major restructuring in Britain providing one template for strategic innovation to shift the established hierarchy of paradigms at all levels from the universities to the primary school. Even the much praised Japanese educational system is under review. In Australia there are novel proposals to tax the future earnings of university studies, and that is a logical deduction from theories of human capital.

The shift within corporations is virtually taking the form of 're-paradigming', with deliberate attempts to shape the consensually accepted categories and grammar of organizational decision-making (see chapter 9) to target accountability as an individual and collective goal. Contem-porary organizational grammar emphasizes flexibility in combination with long-term strategies. This is very obvious in the British university sector where the state aims to remove tenure and to separate the traditional coupling of research and teaching in all except the most successful research-based universities. Certain British universities acquired new senior executives and these have acted vigorously to remove weak departments from view and to highlight strengths. It is mainly those universities whose strategic intents emphasized new knowledge that have survived best.

There is little doubting that most enterprises are in the throes of planned changes. Moreover, Clark and Starkey (1988) observe that the pace of corporate inertia varies between enterprises, and within enter-prises varies between the dimensions of corporate strategy. Within enterprises it seems that finance capital can be readily redeployed and that switches of markets are feasible. Also, the ablation of long-established competences seems to have proceeded at a fast pace in British manufacturing and distribution. Thus the problem of structural inertia seems to be partly resolved by switches of direction, by ablation, by joint ventures, by disaggregating vertically integrated enterprises, as well as by the founding of new enterprises. These signs of strategic intent are significant and they are accompanied by novel notions of successful organizations (for example, chaos as a desirable situation). Moreover, the use of finance as the control of internal resources is highly signifi-cant as an adaptive mechanism to simulate externally-based resource dependency. It may therefore be argued that, whilst the population ecology model provided a much needed challenge to the dominant pre-occupations of organization analyses, the strict interpretation is too pessimistic. That stated, it must be emphasized that the strict population ecology perspective has a great deal of relevance to strategic innovation.

Case: the American fast-food sector

Introduction

This section is a case illustration based on the fast-food sector and the growth of McDonald's.

Today McDonald's is rated as the world's fourth largest retailer, America's largest corporate estate owner, and has more than 9,000 outlets world wide. McDonald's is a complex structuring with a core of franchises and some 2,000 suppliers. Half the American population live within a three-minute drive of an outlet, and 96 per cent visit at least once per year as they consume one-third of all the hamburgers sold and one-quarter of all the french fries. In the USA there are 500,000 jobs and more than 8 million people received their initiation into employment through McDonald's. Their training organization now replaces the US navy as the major training organization.

The emergence of McDonald's and its position as a leader amongst the new population of fast-food outlets has occurred over the past five decades. The emergence of this population reflects the agency of many individuals who found and articulated a structural potential for fast food and then developed a vast interfirm network around a co-ordination mode which fuses the market and the organization into a franchise relation-ship. Love's (1986) insightful account of McDonald's examines the opportunities available to the franchises to intrapreneur – that is, to introduce new products from, whilst working within, the franchise system. Yet the rise of McDonald's reveals the significance of the over-arching framework of values and authority which establishes uniformity and combines common economic interests with self interest.

Our illustration commences with the context of the emergence of the fast-food sector in the USA.

Emergence of the fast-food sector

The rise of the fast-food sector is closely coupled with the history of the automobile and with the willingness of the American consumer to utilize faster and more standardized forms of economic activity. The fast-food sector commenced with the innovation of taking orders from customers in their cars and then delivering the food to the waiting passengers. That pattern emerged around 1920, and the earliest known version of fast-food began with a steam-fried, onion-laden burger sold for five cents (in 1921). Many small, roadside operations surfaced, but were soon outpaced by chains like Howard Johnson.

The main growth occurred during the 1930s in California with the introduction of open-air parking lots around a central area at which meat was cooked, hand sliced and placed in sandwiches the carhop. In California's favourable sunny climate there was a massive growth of carhops, yet their template of operation remained very similar to that of the conventional restaurant, especially in the length and choice from the menu as well as in the methods of cooking and preparing the sandwiches and in their distribution to the consumer. One innovation was to transform waiters into bellhops on skates to speed the delivery (see 'American Graffiti'). Even in the 1930s the methods of operating were subjected to film analysis, and training films were constructed for distribution. By the late 1930s there were many entrepreneurially-owned, local outlets and there was the beginning of a small degree of franchising by the successful names. The early franchises were loosely controlled. Then in the early 1940s Bob Wian began to open multiple sites under the name of 'Big Boy'. At that time the typical drive-in sold cheap, fast food to noisy teenage groups. There was intense competition between the drive-ins.

McDonald's I

In 1937 the two McDonald brothers – who were fascinated by the problems of speeding up service whilst maintaining quality – opened a tiny drive-in in California and 'stumbled on the cutting edge of the fast food service' (Love, 1986: 12). At that stage their drive-in followed the normal pattern of having a central barbecue pit for preparing the food. Later that pattern was to be altered. Three years later they carried their learning from this experience to a new site and opened up a drive-in at San Bernadino with a central octagonal building of 600 square feet. The building had a slanted roof and the kitchen activities were entirely exposed to the vision of the customers. The menu only consisted of twenty-five items which focused customer choice and the problems of serving. There were twenty carhops who were able to service 125 cars at a period. Although the approach of the McDonald brothers stayed with the drive-in template they made small innovations like adding bar stools at the front of the counter facing the kitchen. Also, their approach emphasized the image of quality for a family unit rather than the teenage gang. The drive-in attracted large numbers, especially of families. Soon the turnover reached 200,000 dollars per annum (1940 values) and their profitability – which became known through trade journalism – was attracting attention to the novel elements of their approach.

McDonald's II

Competition increased and the brothers considered the alternatives for maintaining their profit levels. They analysed all their sales receipts and discovered that 80 per cent of the sales had been for hamburgers. So, in 1948 they closed their business for three months to plan a new approach. During the break an array of innovations were introduced which included replacing the bellhops with a service window standing in front of a completely redesigned kitchen layout with new raw materials, equipment, products, and organizational practices: a new configuration. The new configuration contained twelve elements:

1. The menu was slashed from twenty-five to nine items, three of which consisted of variants on a 15 cent hamburger (ten hamburgers to the pound weight) plus three soft drinks in two sizes.
2. The market segment was defined as the family, especially the working-class family.
3. The introduction of an attractive, attention-grabbing presentation for the queuers, particularly the children, who were watching the preparation of the food.
4. The introduction of a paper service (that is, cups, plates) which removed the requirement for washing the dishes and increased the pace of operations.
5. The amount of capital required to enter the fast-food sector was reduced to be about one-third of the previous level.
6. The kitchen was massively redesigned to incorporate shining, easy to clean, steel counters. The area was 12 feet by 16 feet and specially-designed equipment – matched to the focused requirements of the new, short menu – were introduced: two 6-foot grills; stainless steel lazy susan to hold twenty-four burgers; shake mixers with ten continuously operating heads.
7. The spotless surfaces were frequently and publicly cleaned.
8. Food was prepared in advance to anticipate the upcoming demand.
9. Rigid operating procedures. The reduced menu increased the repetition, reduced the need for specialist cooks, and increased the number of quickly-learned tasks.
10. A new approach to the division of labour based on pre-defined and tightly synchronized teamwork. All tasks were analysed to decide their location and their crewing. A twelve-man crew was established with three countermen, two shakemen, three grillmen, and two dressers; two on french fries.

11. Service was based on the 30-second serving cycle in order to sustain the willingness of the customers to queue.
12. Labour requirements were reduced to one-third. The bellhops were no longer required because the customers fetched their own food.

After six uncertain months of operating, the previous levels of turnover were achieved and passed. By 1952 the new McDonald's were the cover story in the American Restaurant Magazine and so the outlet attracted many enquiries and close scrutiny. A new template of operations had been established, but how would this prototype be diffused? How would its success influence rivals? The brothers began to license the concept of the red and white rectangle with a sharply slanted roof, and opened a prototype at Phoenix, Arizona. Their approach to franchising was quite loose and involved selling the rights for whole territories. In practice this method gave away control to the franchisee. However, by this stage a new actor – Kroc – entered the scene.

Kroc's franchising: origins of the modern McDonald's

Franchising began in manufacturing and normally involved the licensing of rights in a defined territory. During the early 1900s distribution networks for soft drinks were franchised. Similar principles applied to autodealers and in the 1930s to the major oil dealers. From the 1930s onward franchises became more visible in retailing with Howard Johnson (1935) and Dairy Queen icecream (1940s). The latter had 2,500 outlets by 1948, all based on the developments in the configuration of elements which make icecream. New equipment was an important element. All these franchises transferred significant areas of control over the operation to the franchisee. In 1954 events began to unfold which led to a new pattern of franchising. The McDonald's template became the vehicle for that new form, yet they were often unwilling accessories to subsequent developments.

The operation of the franchises was of considerable interest to the suppliers of equipment and they were in an excellent position to overview the quickening of the service cycle. Enter Kroc a senior and experienced employee in a large equipment supplier. He reflected on his observations of the franchisees based on his sales visits. He noted certain flaws in the existing method of franchising:

- that the franchisees often exploited the original concept by substituting inferior raw materials in return for kick-backs from the suppliers;
- also commissions were based on the licensing of equipment rather than on the market performance so the central franchiser approached the franchisee from the suppliers perspective;
- that mistakes were magnified and could not be readily corrected by the licensor.

Kroc decided that two significant qualities were missing: uniformity coupled to quality. These could only be achieved by the licensor retaining control of the franchisee. How could that be achieved in a manner which the franchisees would find accountable and would perceive as legitimate?

Kroc set out to construct an organizational system in which the interests of the franchisees were recognized and mobilized, yet the licensor could impose high quality and uniformity. He approached the McDonald brothers at a time when they were experiencing problems with their own methods of franchising and persuaded them to give him the key position to control (and have a stake in) future developments. Under Kroc the new franchising policy gradually evolved, often through trial and error, frequently through chance. Kroc decided to stop franchising territories and only to sell a single store as a test of the capabilities of the franchisee. Further, he made the basis of payment rest on the market share, thereby combining the interests of the licensor and franchisee on the same target. Kroc also resisted immediate, big profits, and set out to establish large sales. Through controls he set out to maintain uniformity and high quality.

This new approach to franchising was not immediately attractive to the existing population of franchisees who preferred to retain entrepreneurial control rather than become involved in a collective bureaucracy. Consequently very slow progress was made, especially in California where it was most difficult to control the franchisees. Moreover a very successful joint venture in the mid-west at Des Moines at a demonstration outlet did not attract much attention. After several unsuccessful ventures the take-off occurred in 1955 about fifty miles from Chicago, where the new store became a social event and focal point which spawned twenty-four further outlets in two years.

At this juncture Kroc, whose patriarchal style focused on quality, service, and cleanliness, hired Sonnerborn. Kroc recruited (and fired) a diverse collection of people – few of whom had college degrees and all of whom were willing to work for what was then a very non-traditional enterprise – and delegated considerable influence to them. Sonnerborn's appointment became very significant, though his initial responsibility was simply to find suitable sites. Sonnerborn interpreted his responsibilities to include the development of real estate and to develop a central bureaucracy. Kroc and Sonnerborn, although often in dispute, were also complementary and Sonnerborn was much more acceptable to the key interface with the bankers who supplied the finance which underpinned the next phases of substantial growth. Sonnerborn also played a part in the crucial development of the corporate infrastructure, including the development of specific techniques and equipment for cooking processes (see Love 1986: 104). Sonnerborn's notion of 'solid and simple' fitted well with Kroc's preference for teaching fundamental principles with which they (the franchisee) could identify.

The new central infrastructure was dedicated to design and development activity on the operating systems, and was a design support to the franchisees. Its development was in striking contrast to the general approach at that time (see Love 1986: 116–26; 132–4; 145). There were four areas of concentration:

1. Improving the product (for example, the milk shake) and the packaging. This included devising a method whereby the franchisee could test the incoming supplies of hamburger for quality.
2. Developing the chain which supplied the raw materials. This included the growing of crops like potatoes. Also included was the development of a close relationship with suppliers to jointly set high standards. The central department modelled future demand and shared the model with the suppliers.
3. Upgrading the buildings and refining the 'equipment package'. The equipment package was tailored to the typical 900-foot kitchen in which fast food was prepared. It included the development of soft-drink dispensers which could supply five times the normal load; new grills; an air-conditioner system; new types of toasters.
4. Organizing a field force to train and advise the franchisees on how to use corporate 'bibles'. Later this became known as the Hamburger University where (in 1983) some $40 million was spent on detailed training of the operators. The field force were responsible for supervising the outlets, and to support their activity the central department devised operating manuals defining the core techniques: 'crammed with minutae' for the franchisee. There was considerable trial and error coupled to great tenacity. Yet the field force of 300 each responsible for eighteen stores was in place by 1958.

Kroc had not charted a plan to make McDonald's profitable to himself and he never used a profit and loss account. Yet the enterprise became successful, especially the real estate.

Sonneborn formed a real estate company to locate and lease sites and, because the franchisee often had few funds, the estate company retained control of the site on a defined lease basis. This method provided McDonald's with an immediate cash flow which was reinvested to purchase real estate. Control over the property also reinforced Kroc's concern to impose quality, service, and cleanliness. Moreover, and by chance, Sonneborn entered the property market at possibly the most propitious moment. Soon an experienced team of professionals developed, so that by 1960 McDonald's had an acknowledged track record in handling real estate.

By 1961 Kroc had a confrontation with the McDonald brothers and he was concerned to extract a larger return for his enormous efforts. With Sonneborn's financial skills Kroc raised the $2.7 million to buy out the brothers. The bitterness of the feelings was reflected when Kroc opened a rival store to the brothers. By 1961 the template of the modern McDonald's was in view.

Part III
Innovation-diffusion

Chapter five

Macro long-term patterns?

Introduction

The search for long-term patterns in innovation has been a pervasive feature for all the approaches to innovation. The approaches taken have been evenly distributed between the stylized facts which typified the quadrant of 'frictionless transitions' and the explanatory and prescriptive models located in the 'innovation and structural repertoires' quadrant. Examples of the former include the early diffusion models of Hagerstrand and Mansfield, and the Utterback–Abernathy framework. Examples of the latter include the Dosi–Freeman frameworks which, because of their basis in the Schumpeterian perspective, give full attention to the structural obstacles and enablers to innovation.

The discovering of patterns in innovation really commenced in the nineteenth century when Marx contended that the economic pulsations of capital operated on a short-wave pattern of about seven years in length with each successive wave representing both a linear extension of the previous direction and also being a source of cumulating rupture and contradiction in the relations between capital and the civil society. Late twentieth-century analysts of macro long-term movements have tended to search for pulsations of around fifty years and to suggest that each successive wave is sufficiently different to the previous wave. Also, twentieth century analysis has shifted uneasily between the heritage from Marx, and the Schumpeterian perspective in which capital as a formation is being constantly renewed.

The revelation of long-term empirical patterns seems to pass through two phases. First there is the claim to have discovered an empirical regularity whose basis in statistical data is portrayed as conclusive, and whose wave-like profiles are evident (see Utterback and Abernathy 1975; Kondratiev 1976). Second, subsequent thinkers – and even the original contributors – qualify the earlier position by suggesting 'the patterns' are principally of heuristic value as guiding principles in the development of a processual, dynamic perspective. Recent contributors tend to adopt the most cautious view.

Three major examples of macro long-term patterns may be cited:

1. The claim to have discovered pairs of upward and downward fifty-year cycles in the movement of prices in basic commodities: the A rhythm and the B rhythm. This perspective has been widely cited in economic history and became an important pillar in the analytic narrative approach of Braudel (1972) whose approach incorporates long-term economic pulsations as one source dynamic.
2. The claim made in the mid-1920s to have discovered upward and downward price movements in basic commodities within a fifty-year period, and their formulation into the so-called Kondratiev long waves. In the 1930s these patterns were later incorporated by Schumpeter (1939) into his theory of business cycles, but were then largely neglected until the 1970s when two different interpretations developed. (Mandel (1972, trans. 1978) and Mensch (1975, trans. 1979) produced tight reinterpretations, whilst Freeman at the Science Policy Research Unit (SPRU) used long waves and Schumpeter's notions of economic dynamics as heuristic organizing frameworks which policy-makers could use, yet do so in a cautious manner.
3. The Utterback–Abernathy model of the technology life cycle for the sector. The Utterback–Abernathy model surfaced in the mid-1970s as a 'hard-framework'. It was soon extended (Abernathy 1978), but then revised (Abernathy, *et al* 1983) and then reformulated in the direction of descriptive frameworks and interpretive analysis (see Clark, K.B. 1985). These more flexible concepts were designed to provide the corporate strategist with guiding themes for interpreting ongoing events, rather than to rely on simple guideposts. This chapter will give particular attention to approaches of Freeman–Dosi and the Utterback–Abernathy model.

The potential for achieving some degree of analytic integration between the long-wave generic model and the Utterback–Abernathy model has been sketched in Freeman's search for the mediations through which institutions and corporate decisions operate at the design level (see Freeman 1983). Moreover, Freeman's analysis reconnects the economic and societal dynamics. Economic systems are allocative and societal systems are authoritative (Giddens 1979). The linkages between the economy and the social structure of society require considerable further development. Within those dynamics the economic sector is a key meeting-ground for the unfolding of the pathways of innovation, for the influence of societal institutions, and for the strategic choices of specific organizations. Freeman and Perez contend that those dynamics can be most fruitfully explicated by transposing the notion of paradigm from the level of scientific analysis to the bodies of knowledge for cognitizing and handling technologies within the firm. The notion of paradigm as applied to technologies is at the centre of current analyses, much of which is social constructionist (for example, Latour 1986; Pinch 1987). Moreover, the notion of paradigm provides a very convenient bridge between our 'innovation configuration', the Freeman–Perez framework, and the revised Utterback–Abernathy framework.

Paradigms and pathways

This section applies the notion of paradigm to the innovation configuration of equipment, practices, raw materials, and the built environment at the level of the organization, and introduces the notion of paradigm pathways. Previous studies have concentrated upon scientific paradigm. In this section the concern is with 'technological paradigms' defined in accordance with the notion of the innovation configuration. As yet there are few studies which reconstruct the innovation paradigms within firms and the interfaces between established paradigms and incoming generic technologies (Dosi 1984). Therefore the examples which are offered are indicative of what is required in future research.

It is reasoned that paradigms follow multiple pathways containing many cul-de-sacs and 'secret gardens', with many alternative routes, and where the signposts may be pointing in the wrong direction. The pathways are evident retrospectively, though at any moment many players will believe that they have an accurate map of the relevant terrain. The pathways are created by intersections between the diverse players who are enmeshed in innovation activities. Pathways are often shaped by hidden institutional factors which mediate the formation of paradigms. It seems likely that within any society there may be a 'hidden' hierarchy of technological and scientific paradigms which reflect and reinforce the institutional core. Changes in that hierarchy of paradigms may only take place over many decades because of the filière of channels which have been created and sedimented (Clark 1987: chapter 7; Maurice, Sylvestre, and Sellier 1986).

There are three main issues: the meaning of the notion of paradigm when applied to innovation in the organization; the notion of paradigm pathway; the impact of societal institutions on the pathway. For simplicity in drawing upon certain authors we shall refer to technological paradigms and these are to be directly equated with the innovation configuration introduced in chapter 3.

First, the notion of paradigm has been transposed from the investigation of scientific to technological paradigms. Attention to the notion of paradigm may be seen as part of the general thrust to examine enterprises as cognitive arenas containing 'thinking practices', logics of action, scripts, forms of discourse and language and situated practices. The intention is to prise open the black box of innovation-design and possibly to discover that it is occupied by Sartrian engineers (Latour 1986). Paradigm may be defined as a heuristic outlook which establishes how a set of puzzles will be conceived and the means through which solutions ought to be sort. A paradigm is a prescription of the positive routes to follow and the negative routes to avoid. Consequently, paradigms have a powerful exclusion effect by limiting the number of considered solutions to any puzzle. It will define the puzzle-solving regime (Nelson and Winter 1982) and will contain a choice of direction which we shall refer to as the pathway. Technological paradigms will be less well articulated than the scientific paradigms reported by Kuhn and others. The definition of the paradigm will be looser, and the distinction between normal and revolutionary activity will be more blurred, and its shifts will probably be more readily impacted by micro-politics, especially by the influence of powerful groups within the enterprise. The paradigm is likely to be influenced by the conception of trade-offs between different alternative branches of development. The technological paradigm consists of know-how, procedures, cumulated experience of success and failure, and will be partly embodied (as in Figure 3.1). The expertise will be highly specific to the firm because applying the notion of paradigm to the level of technology rather than science introduces certain nuances. It may be presumed that inside certain firms there is an expertise which is coupled with patents, technological know-how, and a network of interfirm linkages (see chapter 8). A paradigm, once constructed in its local context, contains the promise of discovering successful solutions to problems by following a given menu of puzzle-solving procedures. Griliches (1957), for example, emphasizes that 'hybrid corn' was the invention of a method of breeding superior corn which was adapted to the exigencies of a specific locality so that the breeding of adaptable hybrids had to be done separately for each locality. Once the paradigm is established it may well develop a strong momentum of its own, though it must be emphasized that this depends on the constellations of politics and groups within an enterprise. Their relative power and intentions will be highly consequential. For example, the Utterback–Abernathy framework is a paradigm within the academic discipline of the study of innovation, and its application by Abernathy (1978) in the study of Ford did over-focus attention in rather precise directions (see later), and for a period did screen out alternative explanations. Typically the users of a paradigm will become 'blind' with respect to certain alternative interpretations of the 'facts' which are available. The technological paradigm will actually delimit the portfolio of problems to be considered, and is a first level of selection which acts as a focusing device for all puzzle-solving activities. It is a prediction of the facts which ought to be uncovered and how they can be interpreted.

The examination of specific technological paradigms at the level of the firm may be seen as both an explication of the organizational and political processes involved in the 'science push' perspective, whilst also unravelling why the 'market pull' analyzes cannot cope with the uncertainty which often surrounds the evolvement of innovation because the meaning of any price signals cannot be satisfactorily anticipated (Nelson and Winter 1977; Dosi 1984). Even in relatively routine situations there will be considerable uncertainty about the trade-offs between following different directions in concentrating technological effort. Car firms have to decide the extent of investment that they will give to developing different facets of the car. For example, Japanese manufacturers have backed the market potential for four-wheel-drive vehicles. These trade-offs will involve mixing choices which possess strong economic and technological dimensions. In particular there will be trade-offs between economies of scale and economies of scope. Firms attempt to create directions in which there is a multidimensional trade-off, yet that is always conceived in a context of uncertainty. It is uncertainty which is pervasive because the embodiment of technological expertise into new products and capabilities produces mutations which encounter the fluctuating selection criteria in the market place.

Second, the notion of paradigm pathway is used to conceptualize the directions (in the plural) taken over time by a paradigm. The notion of a pathway is preferred to that of a trajectory because the latter invokes an over-precise image. The pathway is reconstructed after the event from the directions taken by problem-solving activities. The pathway will contain many branches, a good proportion of which lead into mazes and into temporary culs-de-sac which cannot be connected until later developments emerge. The pathway will be an outcome of many interacting factors and hence apparently similar systems will take different routes. The classic example is the difference between Chicago and London in the pathway of evolvement of public electricity suppliers (see Hughes 1983). Pathways will vary in the degree to which they are narrow or broad. Pathways reflect cumulative efforts and so will possess a highish degree of momentum in certain directions. That means that switching from the existing paradigm (that is, exnovation) to another paradigm can be extremely difficult and also expensive.

We should distinguish between the reconstructed pathway of a generic innovation for a sector like semi-conductors from the pathways taken by individual firms in that sector. The pathways of firms in the same sector may well contain more variations then previous accounts have suggested. Paradigm pathways are the outcome of multiple firm-specific approaches to innovation (Pavitt 1984, 1987). Moreover, because technological paradigms contain such a high proportion of tacit, non-verbal knowledges, their subtleties may be both hard to detect and yet immensely consequential to the exploitation of strategic edges (Porter 1985). Dosi's (1984) examination of differences between firms in the semi-conductor industry reveals sharp and fluctuating shifts in the relative degree with which certain firms were at the robust frontier of the paradigm. For example, the British firm of Lucas invested in the new technology at an early period, yet seemed unable to gain significant benefits. A parallel example would be of the investment by the speciality car firm Rover in the gas turbine engine. In each case the firms were heading down culs-de-sac although that interpretation is only apparent *ex ante*. The notion of paradigm pathways highlights the significance of firm specific knowledges, and throws an important light on the relevance of various approaches to documenting corporate paradigms. Technological paradigms are rarely studied. Many more studies are required and these should include the impacts, direct and indirect, of societal institutions.

Third, the impact of external institutions on paradigm pathways has been the subject of much speculation, especially with respect to the impacts of the military in the USA and in Britain on the whole fabric of civil developments. Dosi (1984) contends that the US military had a dramatic impact upon the evolvement of semi-conductors through the enormity of the investment and their capacity to create both a diverse array of alternative tests (that is, variations or mutations) on the same puzzle domain, and also in the creation of subtle bridging institutions to orchestrate the cumulative direction. Given the high level of uncertainty, the military occupied a key co-ordinating role in mapping branches, and in selecting fruitful pathways, as well as in cumulating experience. Noble's (1984) essay on the influence of the military on the 'factory of the future' reinforces the uncertainty dimension and highlights the complex concentration of struggles and conflicts between contending groups.

The relevance of the institutional impact on technological paradigms operates in at least two dimensions which are worth following. On the one hand there will be mediated impacts on 'early filtering parameters' which become internalized. For example, it seems that the financial largesse of the US military's support for computer numerical control machines may have skewed the basic design parameters in the firms which supplied the military (and their suppliers) away from medium-term considerations of cost effectiveness of computer numerical control for the civilian user, thereby creating an opening which was entered by the Japanese suppliers at a later stage. The early filtering parameters are a 'mutation generating mechanism' (Dosi 1982) whose scope might be broadened or narrowed by the actions of the state.

We suggest that the institutional core of any society probably incorporates a 'hidden hierarchy of past paradigms' whose influence may be consequential on future adaptation depending on the future orientation of the society and its particular enterprises. For instance, it is possible that the relative strength of British science at the invention state, and its influence on the drug industry, and on the chemical industry, reflects the pro-active state actions of the 1870s in seeking to superimpose the model of German science on the higher education sector. The hidden hierarchy probably influences the formation of new knowledge and may explain the particular role of the social sciences and of economics in Britain. It may also influence the unfolding of new long waves, possibly by the provision of certain kinds of expertise and infrastructure. Schumpeter contended that an incoming long wave required the 'creative destruction' of many areas of social practices and that would include the existing hierarchy of paradigms.

Freeman and Perez suggest that the unfolding of any incoming generic innovations is heavily shaped by the pre-existing institutional fabric of societies. These institutional cores have varying effects. Some cores may facilitate the incoming paradigms because the degree of friction caused by existing structural repertoires is very slight and might even be enabling. That situation might apply to the introduction of information technology, biotechnology and advanced manufacturing technology in Japan. Also, some states might take actions which mobilize the removal of blockages and hence facilitate the creative destruction of the paradigm hierarchy created by pre-existing paradigms (for example, Thatcherist deregulation).

Long waves and generic innovations

There are two major contemporary approaches to long waves. On the one hand there are those like Mandel (1978) and Mensch (1979) who have passionately espoused the relevance of a long-wave perspective which they have based on modifications to the original contributions from van Gelderen and Kondratiev, and there are those like Freeman and Perez who are intrigued by the heuristic value to policy-makers of situating everyday observations within a larger, more extensive framework of analysis. This chapter will concentrate upon the Freeman-Perez notion of long waves as techno-economic paradigms.

The existence of long waves is hotly disputed because the data on which the claims have been made are considered to be too contentious. Because each wave lasts between forty-five and fifty years, only three complete waves could have occurred between 1800 and the mid-1920s, and only four waves by the 1990s. Most contemporary discussion reverts to the data and interpretation by Kondratiev (1925, trans. 1976), whose long waves consisted of almost equal periods of twenty-three to

Table 5.1 Long waves

	UPSWING	PEAK	DOWNSWING	ORIGINS
1.	1793–1825	1825		**COTTON IRON** Expansion of world markets
			1825–1847	Competition between England and Western Europe reduces profits. Rate of expansion of world market decreases
2.	1848–1873	1873		**RAILWAYS STEEL** Growing industrialisation and massive expansion of world market. Railways constructed in Europe and North America. Introduction of machine-made machines
			1873–1893	Profits fall. Diffusion of transformation machines reduces profit levels
3.	1894–1913	1913		**ELECTRICALS CHEMICALS AUTOMOBILE OIL REFINING** Extension of world market. Slow rise in price of raw materials
			1914–1939(45)	War disrupts world trade
4.	1940(45)–1966	1966		**ELECTRONICS AEROINDUSTRY** Intensification of the international division of labour
			1967–?	Falling levels of profit because of intensive international competition

Source: Mandel (1978).

twenty-five years of upward and downward movements in prices. It seemed that these long-term price movements embraced the seven-year business cycles reported by Marx, and so the long-wave theories implied that capitalism was subject to periodic pulsations rather than linear implosion. Kondratiev's interpretation was viewed as a criticism of Marx. The closest modern work to that of Kondratiev is by Mandel and Mensch. Mandel argues that each long wave is set in motion by the swarming of the commercial exploitation of a whole array of inventions. Four long waves are identified as shown in Table 5.1. It may be noted that Mandel identified 1967 as a turning point in the post-1945 era of upward economic expansion. According to Mandel the periodicity is very stable. A similar point was made by Mensch (1979), who argued that 1984 should have been the year in which there were evident signs of the start of a new period of innovation which would slowly accelerate into the 1990s.

Freeman and Perez adopt a more heuristic and tentative usage of *K* waves and prefer the notion of a 'major structural crisis of adjustment'. These adjustments vary in their form, timing and intensity (Freeman 1982: 26). The perspective follows Schumpeter in rejecting equilibrium models (Freeman 1986) and in seeking to elaborate the theoretical requirements for a dynamic model. The studies and interpretation are focused on the search for patterns in generic innovations like information technology, biotechnology, advanced manufacturing systems, and raw materials technology. All innovations are conceptualized as paradigms with many potential pathways and branches (Nelson and Winter 1977; Dosi 1982, 1984; Perez 1983). These generic innovations seem to emerge at irregular intervals which are best considered in terms of event time rather than calendar time. Moreover, several different generic technologies might surface at approximately the same calendar time and appear to coincide. These generic technologies are areas where scientific invention is transformed into commercially exploitable technologies. A generic technology, once established, may create a new sector and also disrupt many existing sectors. A key issue concerns their paths of diffusion because these will be uneven in their appearance in particular sectors. That unevenness reflects differences in the capacities of institutional cores and organizational members to facilitate the incoming innovation, and to handle the problems of exnovating existing practices and devices.

There are three elements in the Freeman-Perez perspective:

1. Capital accumulation is central and is affected by the expectations of the investors and the general rates of interest on capital. Expectations are assumed to be quite volatile so that tendencies towards small booms and slumps are exaggerated. These fluctuating expectations will be easily stimulated by hopes of rapid

Table 5.2 New technology systems

	'Main carrier' Long wave			
New technology systems	Previous long wave	Recovery & boom	Stagflation	Depression
RESEARCH INVENTION	Basic invention/science coupled to technical exploitation. Key patents. Many prototypes. Early basic innovations	Intensive applied R&D for new products & applications. Families of related basic innovations	Continuing high levels of research activity. Emphasis shifting to cost saving. Basic process improvements are sought	R&D costs exceed value of investment. Slackening growth of sales. Yet process innovations attractive
DESIGN	Imaginative leaps. Rapid changes. No standardization, competing design philosphies. Some disasters	Still big new developments but increasing role of standardization & regulation	Technical change still rapid but increasing emphasis on cost & standard components	Cumulative, routine, minor improvements of cumulative importance
PRODUCTION	One-off experimental & moving to small batch. Close link with R&D, & design. Negligible scale economies	Move to larger batches & where applicable flow processes & mass production. Economies of scale begin to be important	Major economies of scale affecting labour & capital, but especially labour. Larger firms	Excess capacity
INVESTMENT	High risk speculative. Small scale. Fairly labour-intensive. Mixture of small & medium firms	Bunching of heavy investment in build-up of new capacity. Band-wagon effects	Initially continuing heavy investment but shifting to rational-ization. Continuing rapid profit growth	Relatively low levels of investment. Underutilization of the capital stock in some of the most modern sectors of the economy: low profit margins

MARKET STRUCTURE DEMAND	Innovator monopolies. Consumer ignorance. Some new small firms to promote basic innovations	Intense technological competition for better design & performance. Falling prices. Big fashion effects. Many new entrants in early build-up	Growing concentration. Intense technological competition & some price competition. Pressure to export & exploit scale economies	Even stronger trend to oligopoly or monopoly structure. Bankruptcies & mergers
LABOUR	Small-scale employment generating effects. High proportion of skilled labour, engineers, & technicians. Training & learning on the job & in R&D	Major employment generating effects as production expands. New training & education facilities set up & expand rapidly. New skills in short supply. Rapid increase in pay	Employment growth slows down, & as capital intensity rises, some jobs become increasingly routine	Employment growth comes to a halt. Unemployment rising. Employment suffers (in the first instance) from the general recessional & depressional tendencies in the economy at large
EMPLOYMENT EFFECTS ON OTHER INDUSTRIES & SERVICES	Negligible, but imaginative engineers, managers, & inventors are thinking about them & planning & investing accordingly	Substantial secondary effects, mainly employment generating but gradually swinging to displacement	Labour displacement effects, as new technology now firmly established & strongly cost reducing	Continuing labour displacement as new technology penetrates remaining industries & services

Source: Simplified from Freeman (1982).

accumulation, so will occasionally result in band-wagon effects. The dynamic role of the entrepreneurial investors will play a major role in allocating financial resources between sectors.
2. At certain times the activity of entrepreneurs will bring together clusters or families of innovations which incorporate radical steps, so technical change will both produce equilibria and will also create certain inflexibilities, lags, and discontinuities. The crucial clusters of technologies are those which permit new markets to grow rapidly so that heavy flows of capital can be raised and invested.
3. If long waves do exist, then the band-wagon effect will stimulate innovations from the outgoing down wave – the previous Kondratiev – whilst also sustaining the emergence of a recovery based on the new carrier wave.

If long waves exist they will produce stylized patterns of evolvement in several facets of the organization: in research and development, in design, in production, in investment, in market structure and demand, in work organization, and in forms of employment. Table 5.2 is a highly schematic interpretation of how a best practice organization should have adapted if it wished to be a general (cf. speciality) producer during the previous four long waves. The schema provides a thoughtful starting point for scenario writing about the future.

Freeman prefers to refine and develop the notions of disequilibrium suggested by Schumpeter, rather than perpetuating what he considers to be the weaknesses in the analysis of technical change by neo-classical and neo-Keynsian theories because these fail to take account of the specifics of changing technologies. In particular, Freeman casts doubt on the relevance of the factor substitution models of innovation proposed by the neo-classical positivists, the reason being that the surfacing of a new techno-economic paradigm introduces a new strategic formulae for the best practice, and creates new sets of rules to be utilized by those most centrally occupied within the techno-structure: designers, engineers, and entrepreneurs. Therefore there is a whole array of rapidly changing production functions for existing and new products. The new paradigm brings an enormous leap in potential productivity. However, the diffusion of the new paradigm will be very uneven through sectors.

The surfacing of several new generic paradigms 'demands' major alterations in existing forms of expertise and in organizational arrangements at all levels: institutions, sectors, and firms. There are, and will be, immense problems of adaptation and the forms of these will affect the pace and pathway of diffusion. The new paradigm enters a world dominated by the old paradigm and its comparative advantage is demonstrated in only a few sectors. The speed of diffusion will be influenced by a combination of falling costs, sustained profits to the

suppliers including the growth of the supply side, and by the demonstration value of the early applications. Consequently there may be
a 'mismatch' between the organizational and institutional requirements of the new paradigm and many of the existing practices (cf.
Perez 1983). The degree of mismatch will vary between firms in
the same sector as well as between societies. The pace and pathways taken by the new paradigm will be influenced by the power
and intentions of already existing social groups. The actions they
take in bargaining, and their ability to confine existing power centres
whilst mobilizing openings for those groupings associated with the
new forms, will shape adaptation at the organizational level.

The Freeman-Perez perspective can be illustrated schematically
to emphasize that generic innovations both create new sectors (for
example, semi-conductor industry) and also have uneven impacts on
existing sectors. The argument would be that the automobile sector was
founded in the third K-wave and then unevenly impacted by successive
waves. If this depiction has any basis then the Utterback–Abernathy
framework examined in the next section requires reformulation in order
to cope with later periods of epochal innovation and their uneven
patterning.

The possibility of long waves draws attention to the variable balance
between the economies of scale and the economies of scope. Earlier
theories presumed that the economies of scale would become more
exacting with the increasing longevity of the sector. However, some
innovations in equipment, in raw materials, in the built environment,
and in organization are permitting a reduction in the economies of scale
and an increasing role for the economies of scope. This issue is central
to the next section.

Finally, an important feature of the long wave perspective is that each
generic technology seems to be anchored in a small number of concentrated spatial locations: regions. The first long wave was initially situated
in certain regions of England, like the 'Black Country', and the nearby
areas of the cotton industry in East Lancashire. The fourth long wave
was based on the USA, and the fifth long wave seems to have important
regional bases in California and in and around Tokyo. This regional
concentration has invited speculation about the spatial patterning: how
it comes about and what might be done for regions which appear to be
unattractive. The notion of a regional technopole has been invoked to
label the concentration of interdependent activities. Also, attempts have
commenced to treat technopoles as instruments of regional economic
policies for innovation.

Life-cycle models: Abernathy revisited

The search for long-term patterns has also been pursued at the sectoral level by Utterback and Abernathy. This section examines the Utterback and Abernathy model and treats the model as an illustration of a paradigm which, once formulated, then sustained the image of frictionless organizational transitions.

The modelling of the long-term dynamics of the sector has been through four iterations in less than fifteen years:

1. The Utterback–Abernathy framework.
2. The 'test' of the framework in a study of Ford from 1900 to the early 1970s (Abernathy 1978).
3. Introducing the notion of de-maturity.
4. The innovation ferment model of Abernathy and K.B. Clark (1985).

The Utterback–Abernathy framework was tested and extended in (2) and (3), then substantially revised in (4).

The first iteration by Utterback and Abernathy was constructed from large-scale data bases which were utilized to proclaim the existence of an empirical regularity in the patterning of the intensity and location of innovative activity. The main framework was illustrated earlier in chapter 1 with Figure 1.1. The Utterback–Abernathy framework stated that the empirical regularity could be used to plan the future allocation of resources, because all sectors moved in a unilinear, irreversible direction, which could be expressed in the metaphor of the life cycle: birth, growth, maturity. Death was not mentioned! Nor was consideration given to the reversal of directions through later periods of epochal innovation arising from new generic innovations.

The second iteration (Abernathy 1978) involved a key test of the analytic potential of the Utterback–Abernathy framework and its translation into a model (see Clark and Starkey 1988: 22–36), and we referred to the broad principles of this analysis in chapter 2 (see Figure 2.6). The development of the model was based on an intensive examination of twenty important, diverse automobile innovations over seventy years. The test was based on excellent archival records of Ford and confirmed the capability of the framework to structure the archival data whilst also revealing an unpredicted reversal of certain production units away from high specificity (see Figure 2.6). This intriguing discovery led Abernathy to emphasize that American car firms had resolved the innovation-efficiency dilemma in favour of efficiency. Consequently there were 'roadblocks to innovation'. Although there were signs of major reversal back towards fluidity in certain productive units, the initial interpretation treated these reversals as perturbations rather than as indicators of the need for significant revisions to the basic paradigm. If we apply the perspective of the 'innovation and structural repertoire' to the first and perspective of the 'innovation and structural repertoire' to the first and second iterations (by Utterback and Abernathy), then there are certain important features left unexplained:

(a) the extensive development of specialized forms of expertise in the technostructure;
(b) the cumulative rigidification of Ford. It was becoming a machine bureaucracy (Halberstram 1986);
(c) oligopolistic massaging of American taste (Clark 1987);
(d) the deliberately slow pacing of innovation;
(e) the absence of systematic comparisons with either Europe or the Japanese firms which were already 'learning by doing' in the American market (Cusumano 1985);
(f) the lack of scrutiny of the transitional problems previously faced by Ford and other US firms in adapting to changed conditions (see Clark, K.B. 1985). There was a failure to situate the automobile industry in its overall context, especially an analytic failure to utilize dialectical thinking in a processual framework. The problems with the model are extensive, yet the vein of concepts and the careful empirical work contributed by this study has been invaluable in clarifying some of the most important requirements of a perspective on 'innovation and structural repertoires'.

The third iteration came in 1980–81 with a paradigm extension eventually labelled 'de-maturity'. This was anchored in an intensive examination of the differences in the approach to innovation between Japanese car firms and the American car firms. Abernathy and his colleagues concentrated upon the detailed unravelling of where and how major Japanese firms had achieved different kinds of innovation, especially the blending of innovations in organization into the uses of equipment and the built environment, yet also stretching outside the firm into its suppliers and the interfirm network (see later: chapter 8). The investigation collated detailed changes to the American structural repertoire at the operating level of 'just-in-case' management control philosophy: rapid change over times, use of 'pull systems' and small batch sizes. Moreover, it became apparent that inside the Japanese technostructure – which differed from that known in America – there was an emphatic approach to design and development on the productionizing of the car and in its design. These provided the Japanese with a very considerable financial saving which was (in 1980) still sustained by the lower exchange rates. These observations 'startled' the observers and led to significant, highly-cited reviews in the Harvard Business Review to diffuse the main interpretation, and then a more extensive analysis of the 'myth' of American manufacturing excellence (Abernathy *et al.* 1983), and an attempt to 're-paradigm' American management by rewriting the popular history of the American system of manufacturing which the British had so admired in the previous century (Rosenberg 1969). In common with other historians of manufacturing it was argued that the American system of manufacturing was itself an ecologically-shaped outcome of the interaction between the predispositions of American consumers and the highly-developed infrastructure of transport and marketing which emerged in the late nineteenth century. The intention of rewriting the history of management was certainly to alter its own self-image and to substantially alter the content of American management teaching away from recipes, check-lists, simple notions of best practice, towards a deeper contextual analysis of factors and processes which had to be adapted to by management. That direction runs counter to large, mainstream themes in American management thinking, whilst providing openings for alternative, more diagnostic and dialectical approaches (Clark and Starkey 1988: chapters 1–4). This policy recommendation shifts the approach of Abernathy and colleagues from 'frictionless innovation' (see chapter 2) into 'innovation and structural repertoires'.

The fourth iteration was the one on which Abernathy was concentrating at the time of his death. Its main features have been presented in chapter 4 in the examination of altering and entrenching innovations (see Clark, K.B. 1985; Abernathy and Clark 1985). Now the invocation to managements is to learn to describe innovations in a detailed manner.

The four iterations provide an intriguing reconstruction of the track taken in developing an analytic framework which is capable of making sense of the current situation in a manner which situates the imperative to action in a more diagnostic and dialectical framework. Contrary to the notion that management can be reduced to a single best practice by following the traits of momentarily successful organizations, the new approach emphasizes the development of a 'thinking corporate culture' in which innovation-design is led by considerable 'front-end intellectual loading'. That is, a great deal of direct diagnosis of situations amongst managements, which share a consensually validated grammar of rich concepts, and cause maps about the environment and about the structural capabilities of an organization: their own and their rivals (see Part IV).

Scale/scope economies

There is an important series of tensions between the Freeman–Perez perspective and that of the Utterback–Abernathy framework which bears very directly on the debate over whether flexible specialization is replacing large volume production. The issue centres on the dynamics between the economies of scale and the economies of scope. This section addresses that dynamic with reference to the automobile industry.

The Utterback–Abernathy framework implied that a number of significant sectors pass through a series of states which crystallize into the state of maturity whereby a small number of very large firms produce a narrow range of product models which are relatively similar in their performance and design and so have to compete on the basis of low price. In the case study of Ford, a single firm vertically integrated the design activity, the supply channels, the assembly process, and also shaped the distribution outlets through contingent contracts. These four areas are being separated to achieve specialization, and being integrated through design-led networks (Miles and Snow 1986; Clark and Starkey 1988). Focal organizations are achieving profitable flexibility through negotiated contracting and implicit control, but without many of the costs of maintaining diverse forms of expertise, and by leaving the subcontractor the responsibility for blending innovations. These developments have highlighted the problems of scope in the product range and flexibility in the production system (Hayes and Wheelwright 1985). They have also challenged the suggestion by Abernathy that there is an inevitable shift from custom production to line production. Instead, the issue becomes one of decoding the dynamics of the economies of scale and scope between custom, speciality, and generalist production, and suggests that the composition of firms which constitute a sectoral network will be varied in respect of small/large (see Rothwell 1986), and in respect of types of production mode (Clark and DeBresson 1989).

Freeman-Perez contend that all sectors will be unevenly subjected to irregular states of discontinuity in which the established innovation configurations will be disrupted through the introduction of new materials and equipment, and that these will combine into new ensembles which impact the built environment and forms of organizing. The notion of irregular periods of restructuring suggests that new techno-economic paradigms are 'superimposed' upon existing sectors like automobiles and textiles and that there are uneven, yet epochal effects. The existence of these impacts has been demonstrated by Tushman and Anderson (1986) for a diverse set of sectors. Their well-documented conclusions are supported by case studies. One implication of the Freeman-Perez thesis is that there is a continuing dynamic between the economies of scale and scope which deserves further analysis.

Economies of scope are achieved when the same firm is enabled to lower the unit price of producing second and further products which are produced alongside an initial product more than if they were produced separately. The economies of scope are probably achieved through the distribution of sharable inputs across several products. Sharable inputs include expertise and equipment. Economies of scope can arise as a by-product of the main product. For example, in sugar-beet refining, the same equipment which transforms the beets into sugar also produces molasses and various ingredients for agricultural fertilizers. These actual and potential economies of scope are considerable. A further possibility exists when the scope consists of diversifying into other businesses. For example, Branson of Virgin began by creating a distribution channel in England for a student magazine which was then used to distribute records and then to distribute new, emerging groups (for example, Mike Oldfield). The expertise which Branson developed in handling the contracts, especially the complex, contingent clauses in the contracts for pop stars, was an invaluable input into the leasing of aircraft at favourable rates to enter the gap left in the transatlantic air routes by the collapse of the Laker airlines. The economies of scope were not addressed in the Abernathy analysis because of the analytic salience given to scale economies.

The scale/scope issue arose continuously in the automobile industry. For example, in Ford after 1926 and it was central to the founding of design in General Motors under the leadership of Sloan and Harley Earl. Moreover, epochal and altering innovations in equipment and raw materials directly affected the ability of smaller car firms to compete. In the monopsonistic American market it was Chrysler who exploited and then were confined by the fluctuating impacts of the investment in research and development (Clark, K.B. 1985: 19). In the European markets – with their very distinctive structure of taste and purchasing capacities – the economies of scale unfolded in a manner which differed so much from North America. Also, in Japan there were sharp differences to the USA. These differences invite attention to the economies of scope and to the issue of whether the economies of scale always meant larger and larger.

The dynamics of scale and scope interactions are neatly revealed in the history of a British firm, Rover, which entered the twentieth century as the proud owner of the patent for the safety cycle, commenced assembling automobiles in 1906, and motor cycles soon after, to become (in 1912) the world's largest firm covering the very wide scope of cars, cycles, and motor cycles. In 1912 the British market was mainly confined to the middle classes, a segment of which purchased the Ford T and gave the kit-assembled branch plant the largest share of the UK market. Rover, the third largest British assembler, attempted to maintain scope within the automobile sector by entering the small car market through acquisition (in 1920), but by 1928 faced exit and had to be rescued by creditors in 1932, after which the firm focused its products into the narrow niche who liked traditional shapes with plenty of wood and leather. Rover had been squeezed out of the newly-formed generalist market occupied by Austin and by Morris into the speciality segment. At that time the typical specialist producer in Europe assembled between 15,000 and 25,000 units across several variants of a core model, and produced these with a traditional small batch production system. Other small firms continued to operate in Britain and in continental Europe in the customized segment. The period of the late 1920s was one when new firms were enabled to enter the speciality market, most notably Volvo and Jaguar.

Thirty years later this speciality segment began to change in scope and in scale. The typical European profile witnessed the burgeoning European market for expensive speciality cars. For example, Volvo shifted rapidly from 25,000 to 200,000 units. Also the German suppliers of Mercedes and BMW were producing larger volumes of speciality cars in ranges. In the 1960s Rover designed and produced the best selling executive car in Britain, yet did so with the volume size of around only 25,000. Rover were acquired by British Leyland (in 1968) who attempted to produce a speciality car to compete in the European market. By this stage the break-even point for speciality car makers had shifted upwards, whilst the economic volume for generalists had fallen in certain respects as a consequence of innovations introduced within firms under pressure from the economies of scale and scope developed for the Japanese market (see chapter 7).

Rover as a marque became embroiled in the most extensive attempt at a quantum leap by a British enterprise – from small speciality into the big league. The firm became a battle ground of competing paradigms: the old guard, many nearing retirement, defending the saga of Rover's rebirth after 1932 versus the new incomers (from Ford) who espoused the notion of a large, volume speciality car. Many problems arose in the design of the car and the factory, yet these were partly overshadowed by the transparent failure to take control of the interface with the market, and with the role of the highly independent distributers. The target of 150K units was never in sight between 1976 and 1985. By then the former British Leyland had been relabelled Rover, and its corporate policy had been shifted away from the generalist into the speciality segments through joint design with Honda. Moreover, by the late 1980s the improvements in Rover's capabilities as an assembler – aided by Honda expertise – had become attractive to a firm with great expertise in new raw materials and new electronics: British Aerospace. An intriguing event for those examining the tensions between the revised Utterback–Abernathy framework and the Freeman-Perez perspective.

The Rover illustration highlights the co-existence and shifting dynamics of the population of firms which constitute any sector. The relative balance between different modes of production – generalist versus specialist – fluctuates and is stimulated by the opportunities flowing from the uneven, long-term clustering of innovations, especially the degree to which they can be treated as entrenching or altering. For Rover many relative innovations which were entrenching for other firms were actually altering.

Chapter six

Diffusion processes

Introduction

The study of the diffusion of innovations was quickly dominated by prescriptions, by 'hard frameworks', and by the deceptively tidy division of analysis between sociologists (for example, Rogers), economists (for example, Mansfield), and geographers (for example, Hagerstrand). Ten basic problems have arisen in the orthodox approaches:

1. There has been excessive reliance on the centre-periphery model of diffusion as the exemplar for all diffusion (see Schon 1971). There has been a marked tendency to refer to 'best practice' in the singular and to assume that this could be discerned by copying the features of successful enterprises in an imitative manner. The centre-periphery model was heavily based on the unique, specific, and ungeneralizable requirements of the American agricultural services, yet the centre-periphery model became the single most-cited model of diffusion (see Rogers 1962, 1983). The model was uncritically transposed from its context to create a simplistic 'how to do it' recipe knowledge.
2. The extensive, though partial, use of the analogy of the epidemic to typify the diffusion of innovations. There was a marked failure to use the analogy rigorously and to search for the mechanisms and channels through which the diffusion might be travelling.
3. The narrow concern with adoption because it was simple to measure sales at a moment in time. Consequently the implementation problems were largely neglected, the specific usages were never examined, and impacts of the innovation on the user were dealt with in too blunt a fashion.
4. The tendency to view the adopter as an individual or to treat the firm as a 'black box' which did not require opening.
5. The extensive reliance on stylized facts like the 'S'-shaped curve, the diffusion hierarchy, and the neighbourhood effects.
6. The excessive reliance on variance models using oversimplified

124

proxy variables for key dimensions and processes (Mohr 1982).

7. The oversimplification of the context of the innovation (cf. David 1975) and the absence of serious conceptual and theoretical analysis of the longitudinal and nested embedded features of innovations (Clark 1987).

8. The neglect of failures and the omission of serious explanation of the causes of failures (see Prunty *et al.* 1987).

9. The treatment of the innovation as static through time and as moving at an even pace through different sectors.

10. The assumption of high certainty about the shape, uses and consequences of the innovation was greatly exaggerated and little notice was given to the fact that many outcomes are only evident *ex post*.

The vision of the diffusion processes provided in the 1970s was highly deceptive.

The study of diffusion seemed to possess a coherent framework of enquiry which was both prescriptive and descriptive. Also there were many hundreds of tight, empirical, research studies. However, it soon became evident that the research studies had produced an almost bewildering variety of statistical associations for the same relationships (see Mohr 1982). Although the paradigmatic hold of the early frameworks continues, the current preoccupation is its replacement. This chapter presents the main lines of current analyses.

It is convenient to commence by taking the original centre-periphery model (Rogers 1962) and then to trace the main lines of its revision by Rogers (1983). The attempt by Rogers and Rogers (1976: chapter 6) to revise the centre-periphery model and apply it to organizations is dealt with in chapter 10.

The centre-periphery model implies that the innovation is marketed through a propagator, but the implications of this and the alternatives have been neglected (Brown 1981). It is important to consider the roles of the suppliers in making innovations available and also to consider whether there are societal propensities to commodify innovations. It may be that Americans are more likely to commodify innovations than are the British. That may reflect the existence of highly-developed techno-structures of college-trained experts and their willingness to explore refinements. Particular attention needs to be given to examining the variability of the supply side and to the consequences for the diffusion process. For example, the case study of the fast-food sector indicated that McDonald's undertook much of the development of the innovations in technology and organization within the firm, and that example poses the issue of the relations between users and suppliers.

Economists have emphasized the significance to the supply side of future expectations of profits and of the role of profits in creating 'corridors' down which innovations are likely to flow (Metcalfe 1981). Moreover, the supply side takes widely-varying forms. One particular form which is briefly noted is the tendency for the supplier to benefit from the learning experiences of the users.

Diffusion models are in the process of considerable revision. This chapter seeks to set out the most relevant elements.

Centre-periphery and imitation: Rogers I

Core features

Despite considerable development and revision, a clear line of continuity of both theoretical and conceptual development can be identified in Rogers's work on innovation from 1962 to 1983 when the third edition of the *Diffusion of Innovations* was published. This is complemented by a second strand which focuses on innovation in organizations. This section is concerned with the former, whilst the latter is addressed later (chapter 10).

In broad terms, Rogers's approach is underpinned by three dimensions. First, it is applied; second, explanation is rooted in the tradition of normative sociology; and third, analysis centres on social process. In the first instance his early work is grounded in the activities of the American agricultural agencies during the 1950s aimed at the promotion of best practice amongst farmers, particularly with reference to the spread of innovations including hybrid seed corn and agricultural equipment and practices. Thus a supply-side emphasis is clear in the approach. Rogers is concerned to explicate the appropriate methods to be employed by central agencies to secure the efficient diffusion of particular ideas, equipment, and practices within targeted communities. This objective is informed by a paradigmatic understanding in which the diffusion of innovations essentially concerns the flow of information from a centre to the periphery. In the second instance Rogers's formulation of the diffusion process is anchored in communication theory and rural sociology. The diffusion process is conceptualized as the spread of a new idea from the source of invention or creation, through various channels of information, to its ultimate users or adopters. The relationship between centre and periphery is conceptualized as objective, in so far as there is no conflict of interests, for example, between suppliers and users, and is thus beneficial to the periphery. In the third instance Rogers seeks to explain diffusion in terms of community-bound socio-psychological processes which are open to persuasion and cultivation. This stands as a corrective to the overly deterministic formulations of neo-classical economics which reduce explanations of diffusion to formal mechanisms like price and profitability. The formulation which Rogers articulates in the 1962 edition is largely consistent throughout his work although it has undergone a number of extensions and revisions. The general framework for analysis is underpinned by the following understanding: an *innovation* is something that is perceived as new by an individual and which is *communicated* from one individual to another in a social system over *time* (for example, Rogers 1983: 35).

Although we have stated that Rogers develops a supply-side model, his work is perhaps best seen as a model of adoption. Adoption is the central focus of attention and the key to the diffusion process. Given that in communication theory the process of innovation diffusion is conceptualized in terms of the communication of information, the concern is primarily with factors which facilitate its effective flow. Further, analysis is explicitly focused at the individual level, and information reception and resistance are seen as crucial determinants of diffusion. It is the point at which individuals within a social system take the decision to adopt or not, and the influences on that decisions which is the central concern of Rogers's theoretical perspective. There are two essential components to this: first, the individual's propensity to adopt, or his/her level of 'innovativeness', and second, the congruence between the existing situation and the perceived characteristics of the innovation. Figure 6.1 illustrates the features of Rogers's early model. It can be seen that there are four major aspects: first, the antecedent context of potential innovators; second, the characteristics of potential adopters; third, the characteristics of innovation; and fourth, the innovation process.

First, it should be noted that at this point Rogers's understanding of antecedent conditions is conceptualized in tightly-bounded terms corresponding to individual psychology and the socio-psychological parochialism of community: that is, antecedent refers to the culturally embedded predisposition of potential adopters. The understanding of the effect of external influences is restricted to a consideration of the flow of information about the innovation via media channels. The diffusion of a particular innovation among a given population is, therefore, subject to the interplay between information flow and the intrinsic characteristics of communities and individuals.

Second, the characteristics of adopters as a function is approached through the concept of innovativeness which is defined as the degree to which an individual is relatively earlier in adopting new ideas than other members of the social system. This is a key concept for Rogers, and it is the basis upon which he constructs his explanation of the process of the diffusion of innovations. It is also a very problematic concept to which we will return later. The concept of 'innovativeness' is used to produce a taxonomy of adopter categories. These are innovators, early adopters, early majority, late majority, and laggards. These categories

Figure 6.1 Rogers's early model

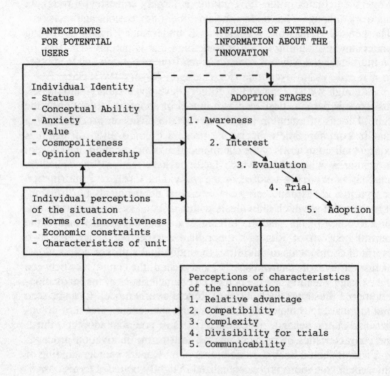

Source: Clark (1987).

are presented as ideal types and their presence in a given community conforms to the normal distribution bell-shaped curve. These adopter ideal types are then ascribed a corresponding set of dominant characteristics: venturesome, respectful, deliberate, sceptical, traditional. A linear scale running from innovators to laggards is then devised to produce adopter types. In this scheme it is held that innovators are more likely to be of high social status and laggards of low social status. Innovators are more likely to be younger, better off, more cosmopolitan, are likely to use more information sources, utilize more specialized operations, be of higher intelligence and greater opinion leadership. Innovators are characterized as social deviants who are 'in step with another drum' (Rogers 1962: 207), who identify with different reference groups outside of the community. The overall schema remains intact, although later versions are more extended and differentiated (for example, Rogers 1983: 260). Specifically, then, Rogers is concerned with varying levels of innovativeness within aggregate populations which he seeks to explain in terms of the socio-psychological characteristics of individuals. His aim is to establish the relationship between these levels and the degree and rate of the diffusion of innovations.

Third, if the level of diffusion is a function of the characteristics of individuals and communities *vis-à-vis* levels of innovativeness, the differential rate of diffusion between different innovations, on the other hand, is explained in terms of the characteristics of the innovations themselves. Rogers produces a list of characteristics by which any innovation may be described and which are of predictive value (for the field agents of the agricultural extension agencies). The five characteristics of innovations are: relative advantage, compatibility, complexity, divisibility and communication. 'Relative advantage' refers to the degree of superiority of the innovation to that which preceded it. Relative advantage may be emphasized by crisis which may either accelerate or retard the rate of adoption. Profitability is an important element in this. 'Compatibility' refers to the degree an innovation is consistent with the values and past experiences of adopters. 'Complexity' refers to the degree to which an innovation may be understood by potential adopters. 'Divisibility' is the degree to which it may be tried on a limited basis, and 'communicability' is the degree to which the results of an innovation may be diffused to others. The way in which potential adopters perceive the characteristics of an innovation is of central importance to the adoption decision.

Fourth, the adoption process itself is conceptualized as a mental process through which an individual passes from first hearing of an innovation to final adoption. This process has five stages or steps. In the early model the stages are 'awareness' which refers to initial exposure to an innovation during which time information is limited. The 'interest' stage involves the individual in a quest for further information. This is followed by a period of 'evaluation' in which the individual makes a mental assessment of the costs and benefits of the innovation. This leads to a small-scale 'trial' before final full-scale 'adoption'. During this process the type of information to which an individual is exposed is crucial. At the early stages impersonal information is important, whilst during the later stages personal information becomes more important.

The rate of adoption within specific communities is a function of the influence of adopters on non-adopters in a social system which leads to the internalization of relative advantage. Opinion leadership is an important concept for Rogers. Opinion leaders are those individuals from whom others seek advice and information; they are significant individuals in the community who by example exert influence on the behaviour of others. They are important elements in the flow of information. This information flow is characterized simply by Rogers in terms of two steps from media to opinion leaders and then to the community at large. The personal influence of the opinion leaders is especially important at the evaluation stage, for relatively late adopters, and in uncertain situations. The concept of opinion leadership leads Rogers to ascribe an important role to the 'change agent' who he describes as a professional person who seeks to influence adoption decisions. They are rarely perceived as credible by their clients who feel they tend to foster the overadoption of new ideas and so the role of the change agent (in this case the agricultural agencies) is the identification of significant individuals (opinion leaders) in order to produce a catalytic effect and generate or accelerate the diffusion process.

As noted above, the rationale underpinning the model is clear. It is intended as an instrument of planning and execution for the use, in this instance, of the agricultural extension agencies in the organized attempt to promote and diffuse their definitions of best practice among farmers. It is worth noting that the model has been enormously influential in the United States in terms of the implementation of both internal and overseas development policy. Nevertheless, despite this paradigmatic success, there are a number of problems with the above model. These concern the concept of innovativeness, the acceptance versus the availability question, the tendency toward pro-innovation bias, and the focus of analysis at the individual level.

First, the most serious conceptual problem concerns the notion of 'innovativeness' which remains largely unchanged throughout this strand of Rogers's work. It is worth looking at this concept more closely for two reasons: first, to illustrate some of its internal limitations, and second, because it acts as a break on theoretical development which has hindered the attempt to overcome the limitations which Rogers himself has recognized in his model. Part of the problem stems from a methodological weakness which is characterized by tautological assumptions. The analysis begins from the empirical observation that adoption rates follow a normal distribution pattern: adopter distributions tend to follow a bell-shaped curve over time and approach normality (Ryan and Gross 1943; Griliches 1957; Rogers 1962: 158). This observation of empirical regularity is useful, but the problem lies in Rogers's generation of categories of adopters ranging from innovators to laggards. This classification is based on ascribed levels of 'innovativeness' but the concept is an ambiguous one. On the one hand it refers to 'the degree to which an individual is relatively earlier to adopt new ideas than other members of his social system' (1962: 159). Thus it is a classification of observed behaviour. On the other hand it is also presented as an intrinsic quality which is possessed to a greater or lesser degree by various individuals. 'One has either more or less innovativeness than others in a social system' (1962: 160). Thus 'innovativeness' is both cause and effect: the incidence of empirical regularity is used to explain itself. In his discussion of the dimensions of his ideal type categories of 'innovativeness', Rogers is sparse with the presentation of evidence. The classifications of psychological types appears arbitrary and he appears to offer little substantial evidence of their actual existence. Rogers (1962) admits that 'there is not unanimous support for this generalisation' (age: 172) and evidence is 'fragmentary' (mental ability: 177). The weakness of this approach is most apparent when variables like 'wealth' are reduced to dimensions of behaviour (175). The problems which are raised by the concept of 'innovativeness' are not simply points of detail but also, and more importantly, concern a serious limitation of scope. In particular, Rogers's preoccupation with this concept is a product of limited concern with what Griliches (1957) has termed the acceptance problem at the expense of the problem of availability.

Second, as noted above, underpinning Rogers's model is the understanding that innovations spread through time and space as part of a process of information dissemination. The spread of information is seen to be even, comprehensive, and innovation is always the best course of action. Thus the focus of analysis is the point of adoption and, in particular, barriers to reception and acceptance. This is why an individual's proclivity to innovate, their 'innovativeness', occupies a conceptual ascendancy. The question which Rogers addresses, to the exclusion of others, is what particular personal qualities and what social attributes produce higher levels of innovativeness in some individuals, and aggregates of individuals, as opposed to others. This is the 'acceptance' question. Within its own terms of reference this approach is useful. It provides a basis for the planning and execution of policy-driven diffusion programmes within specifically selected populations. However, such a formulation is too narrow to contain the diffusion process. What it rules out of court is the 'availability' question. This refers to the activities of supply-side agencies which exert pressure on the process of information and innovation spread through time and space, with the result that not all individuals have equal access to information and innovations. Thus the underlying explanation for variance in adoption rates may be that innovations were not available as opposed to particular individuals or populations possessing low proclivity to innovate. As Griliches (1957) points out, it does not make sense to blame potential adopters for being slower in acceptance than others if particular innovations are not available to them until a later date. In this view it is clear that a great deal of the variance which forms the basis of Rogers's classifications can be explained by variables outside of the individual at an institutional level.

This requires a conceptual shift from the nature and activities of individual adopters to those of diffusion agencies.

Third, the question of pro-innovation bias concerns the implications in diffusion research that an innovation should be diffused and adopted by all members of a social system, that it should be diffused more rapidly, and, that the innovation should neither be modified nor rejected. Thus the decision to adopt an innovation is always seen positively, and the decision not to adopt is seen negatively, the outcome, not of rational choice, but of resistances to innovation. This kind of view has been referred to as the 'technological fix' which refers to an overdependence on technological innovation to solve complicated social problems. In this view technological change is uncritically linked with improvement. This tendency is graphically illustrated in Rogers (1962) treatment of the deterministic relationship between 'traditional' and 'modern' social systems and aggregate levels of innovativeness, with the clear implication that innovation is synonymous with modernity and progress and conversely failure to innovate is associated with backwardness.

Fourth, the focus on individual perception has greatly limited the scope of the model and, in particular, with respect to the analysis of the implementation of organizational innovations. Rogers (1971, 1983) has attempted to address this issue and is himself critical of approaches (which must include his own) which take the individual as the unit of analysis and argues for a need to focus upon social relations. The overarching approach of this later work continues to be one which is centrally concerned with information flows, but there is an awareness that such flows cannot be understood in the earlier communication model and must take account, for example, of the 'sociometric diad' or interpersonal networks which are obscured by the methodological procedure which draws on random samples of individuals. This has the effect of obscuring influential social networks and tends to lead to a poor 'aggregate psychology' (1971: 80). It is pointed out that it has erroneously been assumed that because individuals were the units of response that individuals were also the units of analysis.

Extensions and revisions

In response to the kind of criticisms outlined above, the original model has undergone a number of extensions, refinements and revisions (for example, Rogers and Shoemaker 1971, Rogers and Rogers 1976, Rogers 1983). The work is extended to incorporate a discussion of the generation of innovations and also the consequences of innovation. The conceptual framework remains pretty much intact but there are a number of elaborations concerning points of detail (see chapter 10). The decision process itself is modified and now consists of the following five stages:

'knowledge' which is similar to the earlier awareness stage; 'persuasion' which is similar to the earlier interest stage; 'decision' which replaces evaluation and adoption; for the reasons outlined below an important development is the inclusion of an 'implementation' stage; 'confirmation' which refers to attempts to reduce innovation dissonance. The earlier trial stage has been dropped.

Innovation attributes are also slightly modified. Divisibility is referred to as 'trialability' in order to include the notion of a psychological trial. 'Communicability' is modified to 'observability' in an attempt to move away from the notion of the adopter as passive receptor as opposed to active agent.

However, three developments stand out. First, the recognition that innovations need to be perceived as related components of a broader technology as opposed to unitary elements. Second, the process of implementation is included in the innovation model. This leads to the important recognition that in the process of adoption innovations are also adapted in the context of specific applications. This Rogers refers to as 'reinvention'. Third, there is an attempt to address the question of innovation in organizations which necessarily involves a shift away from a focus on individuals.

Rogers uses the notion of 'innovation bundles' (Rogers and Shoemaker, 1971) or 'technology clusters' (Rogers, 1983) to draw attention to the problems associated with a concentration on single innovations. Favourable perception of innovation is in part a function of compatibility, and this in turn is related to innovations which have already been adopted. Thus further innovation will be dependent upon the relationship of an innovation to other innovations, or components of a technology, which potential adopters are using. There is a need, therefore, to study 'bundles' or 'packages' of innovations rather than each innovation as a discrete and separate unit of analysis. To some extent this tacitly anticipates Gille's (1978) notion of the 'technical system'.

A valuable outcome of this reorientation is the recognition that the study of innovation (diffusion) needs to be extended to incorporate the process of implementation in order to explicate the underlying rationale of decisions to adopt, reject, discontinue, or modify an innovation, instead of smothering this complex process in a blanket of biased assumptions. Further, the incorporation of implementation into the model is important because it most clearly marks a shift away from simple imitation models of innovation which present the adopter as a passive receptor of predefined technologies and/or practices, as opposed to an active participant in the development of technological and social change. Reinvention refers to the fact that, in the process of implementation, innovations are modified, or recombined, in the context of specific applications. This is an important development in the understanding of the innovation process which the earlier simple dichotomy of invention and adoption failed to comprehend.

The question of innovation in organizations is an important development in Rogers's work which entails a shift away from the focus on individuals and towards processes of collective and authority decision-making. However, it receives uneven treatment. Part of the problem lies in the attempt to integrate innovation in organizations into the main body of work on diffusion outlined above which, as noted, is rooted in communication theory and rural sociology. This has hindered the necessary conceptual shift as essentially it still relies on core concepts of innovativeness and information receptance and resistance. However, a second strand of Rogers's work (Rogers and Rogers 1976) in collaboration with J.D. Eveland has made greater progress in this direction. This issue will be dealt with more fully in chapter 10.

Marketing and infrastructure models

The marketing and infrastructure model (Brown 1981) approaches the issues of marketing from a perspective which combines elements of spatial geography and consumer marketing with some features of the economic history of diffusion. Attention is focused on how innovations can be marketed.

The central element in Brown's approach is whether there is or is not an external agency which is capable and willing to play two key roles in the diffusion process. First, the external agency provides a point of distribution for services and/or products/processes in a given spatial area. Therefore, the temporal sequencing and the spatial location of the points of distribution are likely to play a major role in determining the general spatial outline of the diffusion corridor. External agencies tend to establish specific ways of presenting the innovation and to develop strategies for diffusion. Some agencies may specialize in either the adoption or the post-adoption implementation processes. For example, the diffusion of managerial innovations may involve consultants playing a brokering role in the implementation processes which may exclude the design states of choosing the specific bundle of elements. For example, the Ollie Wight agency specializes in providing the educational input to a firm intending to purchase equipment and software for computer-aided production management.

The marketing and infrastructure perspective concentrates upon the description and prescription of the roles which a diffusion agency can and should play. The main framework is presented in Figure 6.2.

Three aspects of the agency are examined:

Figure 6.2 Market infrastructure model

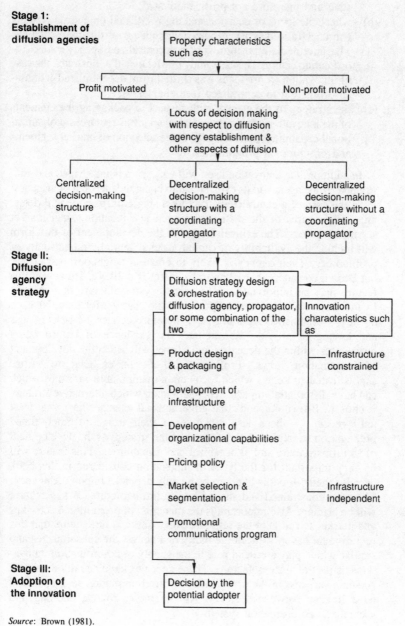

**Stage 1:
Establishment of
diffusion agencies**

Property characteristics
such as

Profit motivated   Non-profit motivated

Locus of decision making
with respect to diffusion
agency establishment &
other aspects of diffusion

Centralized
decision-making
structure

Decentralized
decision-making
structure with a
coordinating
propagator

Decentralized
decision-making
structure without a
coordinating
propagator

**Stage II:
Diffusion
agency
strategy**

Diffusion strategy design
& orchestration by
diffusion agency, propagator,
or some combination of the
two

Innovation
characteristics such
as

- Product design
  & packaging

- Development of
  infrastructure

- Development of
  organizational capabilities

- Pricing policy

- Market selection &
  segmentation

- Promotional
  communications program

- Infrastructure
  constrained

- Infrastructure
  independent

**Stage III:
Adoption of
the innovation**

Decision by the
potential adopter

Source: Brown (1981).

135

(a) whether it is profit motivated or is in some way supported by the state and therefore not profit motivated;

(b) where the locus of decision-making is situated on a scale running from centralized to decentralized without a co-ordinating structure. The intermediate position is of a decentralized agency with a co-ordinating centre. The case study of McDonald's illustrated the case of innovation whose locus was shifted from decentralized-without-co-ordination to centralized (see chapter 4);

(c) the strategy of the agency with respect to packaging the elements of the innovation, the development of an infrastructure and organizational capabilities; pricing policy; the selection of market segments and promotional programmes.

In addition, the innovation itself will vary from being shaped and constrained by the agency in its evolvement to being independent of agencies.

In the case of a centralized structure a single propagator will determine all features of the distribution agencies: location, size, market segments, prices. The critical factors in the development of this form will include the availability of capital, sales potential, and elasticity of profitability. Their agency is likely to create a neighbourhood effect, but Brown considers that a hierarchy effect is unlikely. However, this feature may be important with respect to systemofacture innovations. In the decentralized position (for example, early McDonald's) each agency is established separately and therefore carries the risks of loss and having to provide the capital. The existence of co-ordinating propagator within the decentralized format will affect the development of information flows. Profit-seeking enterprises tend to utilize sophisticated strategies which focus upon manipulable variables which can have differential impacts dependent upon which alternative strategy is chosen. Brown contends that governmental agencies have – at least until recently – been less effective in their usage of finely-tuned propagation strategies. A key feature of the strategy is the development of an infrastructure and of organizational capabilities. This feature will be very important for the types of innovation considered in this book because their format is likely to be heavily dependent upon the agency. They are constrained and shaped by the infrastructure. A key feature will be pricing. Also important is the matching of promotional activities and market selection to the infrastructure. There is little doubt that this requirement has influenced the mergers amongst the consultancies and similar which play a crucial role in the supply of co-ordination innovations as part of systemofacture. These agencies have already displayed a keen awareness of the profitability of different market segments, and have linkage between segments (for example, routine auditing and specialized co-ordination systems).

The formulation of the innovation into a bundle or package is a key consideration. For example, the diffusion of American forms of plant-wide productivity schemes to Britain from the USA was influenced by the skilful packaging of the American concept for the contextual problems of the British market where accountancy data was often less available (Clark 1972b: chapter 9). In the case of large-scale innovations (cf. consumer goods) the innovation may require unbundling and then fine tuning to specific situations.

There are three limitations which require mentioning at this stage. First, despite disclaimers, the approach is more influenced by consumer marketing (for example, ice cream, cable TV, credit cards) than by industrial marketing. So the role of agencies which are internal to large corporations and similar (for example the British National Health Service) is neglected. This is an important limitation in the diffusion of corporate practices because many large enterprises possess sub-systems whose specific activities include the internal diffusion of a collection of innova-tions. Second, the position of innovations which are not handled by specific agencies is largely neglected. Third, there is a general assump-tion that agencies will appear to handle innovations. However, the evolve-ment of innovation agencies may vary significantly between countries. It has been argued that diffusion agencies have been more likely to appear in North America than in Britain (Clark 1987).

Supplier-user interactions

There are two major sets of issues to be considered: the extent to which the suppliers' role is influenced by the search for profits; the attempts of suppliers to appropriate benefits arising from the learning of the user.

First, one of the problems with the centre-periphery model was that it precluded the examination of the economic interests of the suppliers and their role in the diffusion process, especially their willingness to modify the innovation as part of competitive marketing. The role of supply factors was neglected and the dynamic interactions between supply, the innovations evolvement, and the demand side were ignored. We now know that the diffusion of generic innovations is uneven between and within industries (Freeman 1981: 18 f.). Economic studies reported that the rate of adoption increases with the profitability to the user; with the small size of the investment required to secure its adoption; tends to be earlier in science-based firms employing experts; with the size of the adopter. However, these studies underestimated the extent to which an innovation evolves through improvements and may fall in relative cost to the later adopters. Also, given that innovations typically require the development of complementary features (for example, infrastruc-tures), these may be more available later rather than earlier. So, rather than a single diffusion curve there might be a series of diffusion curves each of which is appropriate to a specific environment. For example, the adoption of micro-electronic technologies for managerial co-ordination might possess different curves for different industries, and might proceed faster amongst retailers and financial services than in medicine. The many different diffusion paths probably reflect evolvements in the innovation and in the adoption environment rather than in the process of learning (Metcalfe 1981). Taken together, these many paths form an envelope of successive curves. The shape of the curve ought to embody post-adoption improvements in the innovation (Davies 1979).

The perceived, anticipated profitability of the innovation to the supplier is an important aspect which requires unravelling. The level of profitability will influence the pace of diffusion and the pace will influence profitability. Each incoming innovation has to compete with existing practices. Thus computer-based methods of attaching financial parameters to the planning of materials and resources (for example, manufacturing resource planning) will compete with manual methods. Also there are often retarding factors at work on the supply side through inelasticities in the supply of skilled expertise, of particular kinds of raw material (for example, software for the foundry industry) and similar items.

The rate of increase in the supply side depends on the sources of profitability in the innovation and that will be influenced by: (a) the price of the innovation; (b) the means for its provision; (c) the price of inputs to create the innovation.

For example, we examine the case of the diffusion of computer-aided production management and compare two hypothetical sectors. One sector contains several large firms with an international status, medium-sized firms and small firms. The other sector is typified by many small firms. It is likely that the former sector will generate linkages with equipment and software suppliers through the agency of large firms within the sector and through the attractiveness to them of the original suppliers. In addition, various brokers such as consulting agencies might enter the situation. In the case of computer-aided production management this might arise from the major accounting consultancies extending their involvement in corporate audits to the related facets of inventory and planning. The large-scale suppliers and users can afford the developmental costs and thereby can establish a 'corridor' of diffusion in that sector which will gradually cascade through the medium and small firms. The availability of financial benefits to the adopters and of profits to the users may so reduce the costs of supply that the large, expensive brokers (for example, consultancies) can reduce their costs. Or their former employees might leave and create low-cost agencies to supply the smaller firms. However, in the sector containing only smallish firms, the potential for a diffusion corridor is more limited. Also the crucial role of the brokers in mediating the supply of the generic innovation to the specific needs of the users and their problems of learning will be much more constrained. Diffusion is always relative, and the diffusion environment is dynamic. It is possible that once the innovation moves beyond the prototype state there will be many improvements in both the innovation itself and in the capability of the users to 'learn by using'. The pace of evolvement and diffusion will influence the rate at which economies of scale occur, and will determine the incidence of bottle-necks in the supply side.

Certain further questions arise about the supply side. Are there, for example, certain firms whose strategic policies will shape the supply side? In what ways is the supply side affected by the actions of the state and of institutions? To what extent and in what ways does advertizing, direct and indirect, influence the distribution processes? What are the detailed capacity problems of the suppliers and how do these problems differ between sectors? To summarize: the profits – anticipated and actualized – to the supplier will influence the supply processes and the corridor of the innovation. It is important to examine the suppliers, their numbers, and capacities, and the price they set. It will also be necessary to examine the long-run pattern of costs because subsequent analysis of that pattern could be very revealing (Metcalfe 1981).

Second, it has become apparent that the more complex, radical early prototype innovations may involve a period of considerable reinvention of the innovation and may reveal key learning issues which are relevant to later users. It is unlikely that the users will become involved in appropriating the benefits from this 'learning-by-doing', yet that learning may have a real commercial value, especially to the supplier. The problem of appropriating the benefits from the utilization of an innovation is a basic issue in the economics of innovation. Hippell (1982a,b, 1983) has drawn attention to the various ways in which the suppliers can target, at the prototype stage of development, those users whose experiences are most likely to have a universal relevance. His approach has become known as the consumer-active paradigm, and has stimulated the development on an alternative paradigm of how the user may become the beneficiary. Foxall (1987) has developed a typology of conditions under which the user may appropriate the benefits. However, as we shall see in chapter 8, on interfirm networks, the degree of user activity in developing commercial innovations may occur through mutual exchanges within chains of organizations rather than through commercial exploitation of benefits to a single supplier.

Chapter seven

The international diffusion of innovations: the Americanization thesis

Introduction

The subject matter of this short chapter is the international diffusion of innovations between advanced economies with respect to the transfer of innovations between three countries: the USA, Britain, and Japan. Previous studies of the transfer of innovations in technology and organizations between societies have been heavily concentrated upon transfers from the USA and other advanced economies to the least developed countries (Clark, N.G. 1985: chapters 7 and 8) and have tended to utilize the centre-periphery model which presumes imitation. The concern of this chapter is with transfers between advanced countries.

Much of the twentieth century has been preoccupied with the extent to which American innovation configurations have been superimposed abroad, deliberately imitated, or have served as the framework against which indigenous developments have evolved: the Americanization thesis. The thesis is significant in its own terms and also as a framework for examining the possibilities for the transfer of Japanese practices. We shall concentrate upon the degrees and forms of American influence in Britain and shall indicate how that relationship can inform the examination of the relative influences of America on Japan and vice versa. Arguably Britain and Japan represent two strikingly different encounters with Americanization.

There are four general problems: the tendency to rely upon the discredited imitation model; the accounts of the Americanization of Britain; the implications of recent research on the degree to which the Japanese have appropriated American innovations and created a Japanese system of organizing; the issue of ascertaining how far the processes of innovation associated with the Japanese system of organizing have actually been adopted and utilized in North America and in Europe.

Examining innovation flows between societies requires the usage of multilevel analysis (see chapter 2) and should eventually lead to the construction and scrutiny of complex causal models explicating the key factors explaining differences between countries. These models should aim to include four features:

1. The role of the filière or matrix of societal institutions like education, the family, and the professions in structuring opportunities and constraints for the unfolding of the innovation pathways (as Maurice, Sellier, and Sylvestre 1986; Clark 1987).
2. The relative predisposition of players in different countries to achieve co-ordination by combining the mechanisms of markets, hierarchies and clans in different ways.
3. The significance of the choice by corporations of their markets. It is likely that heterogeneity and pace of saturation of the market differs considerably. Also the heterogeneity and pace can shape organization learning (see chapter 9). These macro-marketing features have contingent affects on the blending of economies in scale/scope which are appropriate and hence influence the survival rates of domestic and foreign firms.
4. The tendency of initial choices of direction (for example, QWERTY) to become established as longer-term learning paths whose directions, which are often shaped by unintended features, become persistent, all embracing and consequential.

These analytic features require development, for example, in highlighting the crucial similarities and differences between the USA, Japan, and Britain.

In this book we can only direct attention to the importance of the societal level and to the role of core institutions. It must be emphasized that the role of core institutions is puzzling. For example, it is likely that many of the accounts of Japan have oversimplified and homogenized the roles of core institutions. So the incisive comparisons between France and Germany by Maurice, Sellier, and Sylvestre (1986) provide a very useful example. The analysis of core institutions implicitly relies on a historical investigation of changing mentalities with the objective of discovering both the typical variety in a society and also to identify those organizations which are atypical and should be recognized as such in international comparisons (Giddens 1985; Clark 1987: chapter 7). For example, the typical variety of British forms of innovation is not immediately made clear by selecting certain well-known firms.

Our aim is to establish an agenda of relevant issues starting with the replacement of imitation models of international diffusion and continuing with a three-way examination of USA, Britain, and Japan.

Imitation, mentalities, and contexts

The problems of the international transfer of innovations were initially subsumed under the centre-periphery model of imitation and principally applied to the transfer from advanced economies to less advanced economies. The western suppliers were not very thorough in examining the problems of transfer for the user – the availability and the acceptance of innovations – though there were some notable attempts to identify and categorize the problems. Consequently, the relevance to managements of the rich vein of studies by economic historians on the transfer of innovations across the North Atlantic was not appreciated until the attempt to increase the transfer of new practices from Japan to the USA and to Western Europe.

The locus of research and theory on the international diffusion has developed mainly with economic historians and with sociologists examining the role of core institutions in national economic performance. The economic historians have made significant contributions, generally through the comparison between the USA and Europe in their relative commitment to technological innovation (for example, Landes 1969), and specifically with respect to comparisons between the UK and the USA. More recently attention has switched to Japan (for example, Rosenberg and Steinmueller 1988). The sociologists, especially in Europe, have undertaken a number of detailed comparisons between countries with respect to their core institutions and have succeeded in penetrating more deeply into the analysis of the structural enablers and constraints within which most organizations have to operate. Often these studies have taken a pair of countries for detailed analysis and for the development of complex causal models linking multiple levels of analysis. The potential was illustrated by: the Anglo-Japanese comparison of Dore (1973); Gallie's (1978) comparison between Britain and France; the Franco-German comparisons by Maurice, Sellier, and Sylvestre (1986). There has been considerable complementarity between the economic historians and the sociologists in their analysis of the causal processes and structures which affect the international diffusion of innovations. Currently these perspectives are being incorporated into macro organization behaviour (Clark 1987).

The limits of the centre-periphery and imitation model of diffusion have been addressed in chapter 6. Imitation cannot be assumed, it has to be accomplished. Rosenberg and Steinmueller lament the failure of American corporations to imitate Japanese practices and it is now recognized that imitation is difficult. Many studies have emphasized the failure to imitate, especially in the transfer of advanced equipment from North America to the Third World. The core issue is of the degree of appropriation of the innovation by the new users. There are several possibilities ranging from total rejection through imitation to total appropriation. The important point is that the analysis of the international diffusion of innovations needs to combine a detailed examination of the innovation configuration before, during, and after the attempts at diffusion, and to do so with respect to both the institutional contexts and to specific users, as well as giving attention to efficacy of the propagation processes.

The Americanization thesis and Britain

Americanization

The Americanization thesis postulates that by the twentieth century the USA had become the most significant location for the introduction and development of new best practice in technology and in organization. Moreover, American best practice was imitated by the rest of the world after a time-lag, the length of which would be expected to vary between countries and between sectors. According to this thesis we should therefore expect to find some combination of four indicators of Americanization in Britain. First, the penetration of key sectors of the British market by American products (capital and consumer) and services (for example consultancy, films, TV). Second, the presence of American firms and their subsidiaries as they pre-empt the emergence of competitive British firms and replace the uncompetitive British firms in a population ecology effect. Third, the presence of American institutional solutions such as their approach to market control and trust busting. Possibly also the presence of American finance capital though that was always unlikely given the low rates of return and the relative strength of the British institution of the City. Fourth, the presence of American inventions like the multidivision form of enterprise, Taylorism, and quality control. How far are those expectations sustained?

The degree of penetration by American best practice will be mediated by the actions of the British in accepting, in developing substitutes, and in rejecting. In turn this mediation will be strongly influenced by the mentalities and historically-shaped practices (Giddens 1985). We examine these aspects in two steps: up to 1945; since 1945.

Britain 1851–1945

Americanization as applied to Britain (and Japan) became an issue after the mid-nineteenth century. It is widely agreed that in the two or three decades following the London Exhibition of 1851 the USA began to surpass Britain as an economic power and that the rate of relative development was much higher in the USA. It became evident that the American exhibits possessed a deliberate, designed, and functional utility. Some products were of a form of manufacture which differed from the normal European practices, most obviously in the key area of gunmaking

(Rosenberg 1969). After the exhibition the British government sent a special committee of three to observe and report on American standard practices of manufacture. Their excellent report stated that American products were typically of functional design, short life-span, cheap, and could be manufactured in large volumes by unskilled operators. Some products consisted of standardized components which were partially interchangeable. Also, the leading Americans used specialized machines and jigs in the manufacturing process. This complex of factors was labelled the American System of Manufacture.

According to Clark (1987) it is highly relevant to note that the attention of the British enquiry of 1853 was directed towards manufacturing because that focus led to a failure to recognize that American success was achieved through the marketing and the distribution which built up high volume demand. For example, American firms continued with European methods of production long after they had used marketing innovations to establish stable, large-scale, standardized demand for their products (Hounshell 1984). This feature is clearly illustrated in the case of Singer sewing-machines, who developed the market for their machines before developing a distinctive American form of large volume production. Likewise Ford developed the demand for his cars with skilful marketing and financing before developing the standardized, moving assembly lines whose development is often attributed to that firm.

In Britain, from 1853 onward, a small group of political leaders from outside manufacturing sought to promote the introduction of American methods of standardizing production and of simplifying the work processes through invitations to American firms and through the specification of government contracts in the direction of increasing standardization of components (for example, arms-making in Birmingham). These nineteenth century interventions by certain members of the government and the civil bureaucracies – which parallel Thatcher's promotion of Japanese methods in the 1980s – were hotly contested, most memorably by the arms manufacturers of Birmingham who considered that American methods created shoddy, non-durable products.

The American penetration of the British market commenced after 1860 and was initially quite slight, mainly because the context and market structure for the usage of American capital products was less favourable. For example, the highly-developed American agricultural equipment was often uneconomic and unsusable on the smaller British fields where its weight might damage the elementary drainage systems. However, the arrival of Singer sewing-machines in Scotland was followed by a steady trickle into various sectors: boot and shoe (1890s), cigarette-making (1900c), automobiles (1912 and 1925), office equipment (1910c) and retail (1910), electricity generation (1920s), electrical capital and consumer goods (1920s). American products – capital and consumer –

became more visible, both through the media and through their actual presence at the place of work (for example, lathes). By 1920 the USA was readily regarded as the source of best practice templates for a vast array of corporate practices. The extent of influence of American best practice in Britain is a significant puzzle because neither the British nor the Americans have fully considered the extent of transatlantic diffusion of innovations and the implications of that experience for the current attempts to learn from Japan. The presence of American firms and of American methods of attaining efficiency and innovation-design steadily cumulated though there was considerable ebb and flow. For example, the highly-organized American lobby for urban electricity utilities secretly acquired key British firms in the London area, but were subsequently forced to withdraw and the supply of electricity became a state monopoly utility. In engineering the American penetration was disrupted by the stock market collapse of 1929. Moreover, the mass-produced American product was not always adapted to the British context and to British tastes. Yet, the American presence cumulated.

From 1890 to 1945 – a key formative period in western capitalism – the impact of American practices of innovation in Britain was probably both hegemonic and disruptive. There were clear efforts by a small number of British firms to selectively appropriate certain analytic practices. For example, two firms in the heartland of traditional, community-based forms of work organization sought to establish 'islands' of medium- and high-volume production in the food industry, and in the automobile supply industry, through the modified adoption of Tayloristic approaches to the analysis of work and its co-ordination; Cadbury's and Lucas. Cadbury's sought to establish work measurement and payment systems through systematic observation rather than the estimating of work times, and applied these principles throughout their chocolate-making process to the wet end and to the dry end of packing. Lucas made considerable and deliberate attempts to observe and utilize American methods of industrial engineering and – relative to many British firms – gave searching attention to the use of jigs at the workplace and to the flows of work through the plant. Retail was another sector in which American methods were studied by specific retail firms. Marks and Spencer, Boots, and other retailers sought to turn back the threatening presence of Woolworths in the British High Street by adapting and translating American stock control and design policies into their operations in the British market.

British responses to the American challenge were often defensive (for example, in the automobile industry), and there was a marked lack of concerted attempts to appropriate those practices and to transform them so that they became available and accessible. Appropriation was often hindered by anti-American sentiments. Also there was the recognition that some features of the American system of manufacturing were not so relevant in the British domestic market which was so stratified and heterogeneous. In Britain the average batch size for the manufacturer and supplier was usually much less than in the USA, and the potential for economies of scale of the kinds achieved in the larger American market with its higher disposable income were absent. The British faced the same problem of achieving economies of scope and scale as were faced later by the Japanese. However, the British started out from a different societal structuration and they constructed a different learning path (Clark 1987).

In Britain there were some attempts to elaborate forms of knowledge relevant to the specific context. The German influence was particularly strong in the new scientific industries associated with chemistry and refining and there was a distinct German influence on the major civic universities which were founded. Yet Germany was not considered as the source of management concepts, nor were management concepts considered significant in a market-oriented society.

Since 1945: France, West Germany, and Britain

There are some interesting contrasts between Britain, France, and West Germany in their response to the salience of American equipment and manufacturing expertise after 1945.

In France a key conduit for the exploration of American best practice was provided, perhaps fortuitously, through the automobile industry where Renault, the major manufacturer, had been transferred from the private sector to state sector as punishment for collaboration with the invading forces. Within Renault a small group of engineers from the political left envisaged a complete package for the rebuilding of the French industry. They recognized that the problem of the transfer of American practices had many dimensions including the market and they analyzed Citroën's failure (in the 1930s) to support mass production assembly techniques because of the problem of demand. The solution was a people's car – the 4CV – which would be mass produced with Fordist concepts and Taylorist working practices in a socialist enterprise (Clark and Windebank 1985). This largely succeeded, especially with the provision of advanced engine assembly equipment in the Marshall Plan of 1948–52. The experience was readily cumulated and edited both within Renault and more widely in the professional associations, and the highly-developed knowledge-making processes of the Grandes Ecoles which possessed a stronger engineering tradition than in Britain.

West Germany might be thought of as the perfect example of the superimposition of American conceptions of market competition between firms and of practices of managing. That assumption has been shown to be mistaken. Contrary to popular stereotypes, the West German economy was damaged far less by the war than had been supposed. There was in 1945 a plentiful supply of cheap skilled labour which was reinforced by the influx from Eastern Europe, and there were sufficient, possibly plentiful, raw materials appropriate to the requirements of the time. Also there was both the continuity of ways of thinking associated with the traditional entrepreneurial élite from the heavy engineering industries of the First Industrial Revolution, and the rising power of the new entrepreneurial élites associated with science-based industries. The former, traditional élites were committed to protectionism and to using cartel-like arrangements to provide high domestic prices. The latter group of élites which came from chemicals, electricals, and automobile production, gradually became more dominant, and their values and attitudes provided a neat bridge between the traditional West German élites and the intentions of the Americans. In some ways the presence of the occupying Americans may have enabled their emergence and their role in shifting the mentalities of the industrial élites on to a uniquely German approach to innovation and to economic development. These existed in some competition between these two élites and their dynamics both sustained and developed a West German economy whilst constraining American attempts to superimpose its forms of market competition. In fact the West German state did intervene to assist a degree of closure in West German markets – contrary to the espoused policy of the occupying American authorities. Also there developed in West Germany a subtle set of relations between capital and labour: each of which was well organized, and both of which shared a commitment to the juridification of relations between employers and employees so that West German workers were much more conscious of their obligations than say, their counterparts in France (Maurice, Sellier, and Sylvestre 1986) and in Britain. So, although the considerable presence of American culture, military might, technological achievement, and life-style had a deep impact, that impact was – as in Britain – considerably mediated by both existing and emerging institutions.

What about the Americanization of Britain? Three levels of analysis should be considered. First, there has been a continuation in the cumulation of an American presence through products, services, and subsidiary firms. Second, from the 1970s onward many large British firms with strong domestic and Commonwealth markets attempted to enter the different markets of North America, of Europe, and of the Pacific Basin. By the late 1980s it was clear that many were experiencing problems, though a small number had been successful (for example, Hanson Trust, ICI). The British seemed to be better at investing finance capital abroad than at the co-ordination of international subsidiaries. Third, in the key areas of innovation and design there was growing evidence that few British enterprises had developed a 'design culture' (see chapter 10). Also it was apparent that the various attempts to learn from the American experience of innovation-design had failed (Clark 1987), for example, in the special joint productivity committees which the American's suggested as part of the Marshall Plan. Also, an array of potentially fruitful American innovations relevant to design had come to litter the landscape of management rather than to become flexible guideposts for skilful adaptation. Management and corporate education were simply an example of these failures. The British educational system never supplied the quantities and types of thinking practices which were likely to provide a knowledge base across firms to complement the firm specific expertise (Clark 1987: 218, 221–6). Consequently, the associations of management and similar which developed in Britain were more social than analytical (see Child 1969). It may be argued that the British did possess institutions which were capable of cumulating experience and translating this into practice, but that these institutions lacked the knowledge-making skills of their international competitors (Clark 1987). So the introduction of the American Masters in Business Administration was a cosmetic superimposed on a vertical, traditional, disciplinary-based, segmented teaching which was largely co-ordinated through the content-free structure of the timetable.

A further issue concerns the persistence of British mentalities and the question of whether there have been transitions which are both uniquely British and also more adapted to the economic context similar to those reported for West Germany. In looking at Britain it is important to recognize the establishment of socialism after 1945, particularly in the creation of state bureaucracies in key areas – health, coal, iron and steel, public utilities, transport – which placed a high proportion of employees in large-scale organization for the first time. Often these new forms possessed techno-structures not found in the capitalist firms which preceded them (for example, National Coal Board), and these technocrats attempted to play an innovatory role in developing and applying new practices. Socialism was largely accepted. An initial step towards introducing tougher market competition was actually inserted into the private consumer sector through the ending of retail price maintenance in the mid-1960s. However, the main shift came after 1980 as the Thatcher government began to pragmatically enhance the role of market forces and directed state intervention to the 'enterprise society'.

There are many subtle, small-scale attempts to draw on elements of the American approach to business, but the main thrust of the Thatcher period has been to reinforce market forces. The problem of providing British organizations with an 'innovation-design culture' has proved awkward. There was a strong persistence of established British mentalities about innovation and about co-ordination – most notably in the Thatcherist belief that the return of market-based forms of co-ordination will be sufficient. As before, the British weakness in devising usable corporate bureaucracies was not analyzed and reversed: instead the problem of organizing for innovation was overlooked.

In Europe the debate over Americanization – which was so hot in the late 1960s – is being replaced by a new debate over Japanization.

An era of Japanization?

Cusumano (1985), in an excellent comparative study of Toyota and Nissan, shows that during the 1950s each firm aimed to learn about car-making from western examples, but that they did so in different ways. Toyota pursued the route of reverse engineering by which European and American cars were disassembled and analysed and some of their features selected for incorporation into the small, very cheap car designed around small engines for the crowded domestic road conditions. Nissan engaged in formal agreements with European car firms which had an experience of making small, medium-volume, low-cost cars (for example, Austin). Cusumano shows that each firm routinely employed executives and experts with a sound graduate education in the sciences and engineering (cf. the British cars industry), and that in addition there was a sprinkling of key persons with experience relevant to the design and development of a car. Both firms developed a collection of recipes for handling the design of the car, the relations with suppliers for the provision of most components, the final assembly which they undertook, and the distribution. At that time there was a minuscule Japanese market occupied by competing motor-cycle firms desiring to edge into the automobile market. It seemed a highly unpromising situation. However, the Japanese car market grew rapidly between 1969 and 1975. Toyota and Nissan were able to adapt flexibly their interorganization networks to the new situation.

Japanese car sales to North America in the mid- and late 1970s were a visible symbol of the close interconnections which were developing between Japan and the USA. Although the Japanese rejected the American incursion of the mid-nineteenth century in favour of European influences (Dore 1973), the presence of American occupying troops and governance after 1945 reflected Japan's crucial geographical role as an island fortress in the Pacific adjacent to the USSR and to China. That role influenced the significant development of containerization for the transport of military material and provided opportunities to undertake various kinds of service tasks in relation to military technologies. By the mid-1970s Japanese cars could be landed on the East Coast for more than a thousand dollars less than the nearest comparable American automobile. How and why?

Cusumano's account is part of a burgeoning growth of articles and books which describe and seek to explain: (a) the basis of contemporary success at a multiplicity of levels – the factory, the infrastructure, the core institutions, the role of the state bureaucracy; (b) the historical evolvement of Japanese entrepreneurship and business acumen; (c) how Japanese practices might be applied in the USA.

In certain respects the Japanese approach to expensive consumer products and to certain capital goods has become the new 'best practice'. Conferences in the west are now devoted to the examination of the Japanization. Describing the details of what the Japanese have been doing and are doing provides a rich, complex picture which contrasts sharply with any analysis of that other island economy: Britain. Explaining the Japanese economic success with innovation is also highly complex. Yet the extensive body of analysis on the British case provides a useful framework, especially for the disentangling of the unintended and the intended.

A sampling from the recent literature on Japanese practices suggests the following ideal type (Smelser 1976):

(a) that Taylorism was imported, unbundled, and applied in its full form to the analysis of shop-floor practices, and the defining of workloads in certain key sectors of the economy. The data from work measurement was utilized to control suppliers, to plan and cost production, to control flows;

(b) that statistical quality control was imported, unbundled, and applied by shop-floor workers as part of everyday tasks. That the objective became one of minimizing defects, and that this is highly effective in final products which contain thousands of components and many subassemblies. Quality control was also linked to consumer acceptability to facilitate product design;

(c) that product design was highly developed as an in-house expertise and coupled to assembly. The final assemblers were part of larger trading groups with considerable bargaining power. The final assemblers often created tight chains of firms whose separate actions were incorporated through contracts, through personal linkages, and by formal linkages in paperwork and in information technology;

(d) the supply of components was devolved to specialist suppliers who were organized, sometimes hierarchically, in a control structure which was shaped by the major assembler (for example, Toyota);

(e) there were many suppliers and they were often incorporated into the design of components;

(f) Japanese firms purchased vast amounts of American invention and basic science through technology agreements. These purchases may reflect the downswing of the previous long wave. The purchased research and development knowledge in forms of patents is readily transferred because of its codification, and because Japanese experts were able to decode the patents;

(g)   the Japanese concentrated upon the design and development usage of the imported knowledges, and in combination with the more tacit, less transferable 'technology knowledges';

(h)   Japan has a quality educational system with strengths in science, mathematics, and engineering up to undergraduate level. There is a larger supply of trained expertise (relative to population) than in western countries, especially Britain;

(i)   a significant proportion of the educated population have perceived their future as involving long-term, service-oriented employment with a single company;

(j)   Japanese corporations have developed their internal education and training in a very extensive manner. They do not compete for the loyalties of graduates;

(k)   firms have actively developed firm specific knowledges, and have actively edited existing paradigms through extensive external intelligence and through some in-house developments;

(l)   leading firms have a benevolent paternalist policy to their management strata, including the provision of apartments;

(m)   management is not segmented and individualized (cf. USA and Britain);

(n)   management involves considerable 'front end intellectual loading' to establish the consensually-grounded grammar of corporate recipes;

(o)   the Japanese save a high proportion of their disposable income and so those funds are available to the banks. Seven of the worlds largest ten banks are Japanese;

(p)   Japanese consumers are heavily concentrated in tight urban environments and living in small spaces (for example, Tokyo), yet cannot invest in house purchase;

(q)   in the past three decades the market has been the largest, most homogeneous and fastest down the learning curve;

(r)   consumers tend to purchase high quality goods, especially those with features of smallness, fidelity in performance (for example, weighing devices);

(s)   the market is highly competitive and therefore batch sizes in that market can be quite small, though that situation has changed considerably since 1965;

(t)   assemblers and suppliers have had to learn how to produce high quality, miniaturized goods in smallish batches and to be able to respond rapidly to either product failure or success;

(u)   managerial culture has retained the highly-developed time-reckoning systems associated with a rural society and combined

151

these with the clock-based systems typical of American managements. The future orientation is highly-developed, multipath, ready for contingencies;

(v)   there has been a degree of focusing of economic activity amongst interdependent complexes, for example, between the development of speciality metals and the automobile industry;

(w)   within firms there has been a tendency to couple the supplier in a tight relationship to the market. Market pulsations and fluctuations are internalized in the 'pull systems'. These features influence the blending of innovation in equipment with other facets of the innovation configuration.

These features constitute an ideal type. Their function is to highlight an array of features.

Although American texts claim that the Japanese firms are simply applying innovations previously invented in the USA, that claim has the same hollow ring as the one made earlier by the British to be the world's leading inventors: who benefits? In fact the Japanese case reveals a great deal of selective appropriation of European practices from 1870 to the 1960s. In the past four decades there has been a strong American influence, yet that has been heavily mediated and contained considerable appropriation rather than imitation. Now the issue has become: to what extent is the Japanese practice the new international best practice? If it is, to what extent can and should these practices be appropriated by European and North American firms, and will appropriation involve major alterations in existing institutions and firms?

Here the British case under Thatcherism is intriguing. If the theories of creative destruction and major economic restructuring have a relevance, then Britain is the prime case of an advanced economy in which major transformations are being inserted, particularly in terms of the exnovation of past practices. Many jobs have disappeared and the existing bases of the remaining occupations are being consciously eroded through both the insertion of market forces and also by the removal of occupational controls (for example, academic tenure). However, the use of market forces by the state apparatus has precedents in the enclosure movement which preceded the so-called industrial revolution when rural workers were dispossessed of their land tenure. Market forces is a characteristic British recipe, and one which has been opposed by the alternative recipe of socialism with its large-scale enterprises. Ironically it was the socialists who did much to develop an innovation culture in areas where private capitalism had failed to do so: the basic industries, the public utilities, and transport, for example. It is the neglect of a design approach to innovation in Britain which cautions against notions of the revival of British-owned manufacturing. Moreover, there are marked weaknesses in Britain's position in those sectors most closely associated with upturns – except for tourism.

The American case reveals much more attention to adapting from large-scale, high-volume speciality marketing in conditions requiring flexibility. We believe that these should not be exaggerated and that many accounts of the transitions have been too optimistic. Yet the potential for adaptation is still well grounded in both large-scale organizations which are attempting transitions (for example, Ford and IBM), as well as the many firms in the speciality sector which are playing a larger role in American exports (for example, Cray and specialist computers). Many of these firms have benefited from the detailed analyses of the Japanese and have sought novel solutions: neither typically American nor imitating the Japanese. That tendency is also evident in the Japanese firms which have been established in North America and in Europe. These reflect a Japanese influence, but not Japanization.

That said, the unfolding of the role of the Pacific Basin in the next century is still happening. The approach we have suggested indicates that its unravelling requires the continued shift of macro organization behaviour away from its roots and routes of the late 1950s into a more descriptive, analytically useful and explanatory approach to the managing of design and innovation.

Part IV

Innovation-design

Chapter eight

Interfirm networks

Introduction

This chapter introduces Part IV, the perspective on innovation-design capability. Part IV is an integral facet of the overall perspective on innovation in technology and organization. It represents a 'chunking' together of analytical problems faced by the users and the generators of innovations. A great many organizations borrow embodied and disembodied knowledge from other organizations, but a relatively small number of organizations occupy the role of major generators and propagators of innovations. Yet all organizations have the daunting task of matching the specifics of their situation to the general features of externally-supplied innovations. Part IV presents and explains the main themes starting with an examination of the roles of interfirm networks in shaping the routes taken by innovations and the consequences of those routes.

The roles of interfirm networks in shaping the generation and dissemination of innovations have become clearer through a series of major research programmes. Three of these will be referred to in this section. The contribution of Swedish researchers to the examination of the innovation-diffusion has been noted through the approach of Hagerstrand, one of the founding authors of that field. More recent studies from the Swedish School have sought to unravel how international firms (for example, Volvo) in small countries (for example, Sweden) acquire, cumulate, and utilize the diverse forms of expertise which are required to be sufficiently innovative to retain a place in the international division of labour. Their longitudinal studies of Swedish firms are highly revealing and indicate how important the appropriate division of labour between organizations is to their collective survival and how linkages between diverse organizations can be orchestrated to handle the 'mechanics of segregation' between firms (Litwak 1961). Canadian studies of innovation have also been inspired by attempts to unravel the relative influence of the context and by a search for the texture of relationships which might exist between different sectors (DeBresson and

Murray 1982). British research has sought to complement earlier American studies in identifying the characteristic forms of innovation generation and dissemination amongst organizations. Who are the suppliers and users? These studies emphasize the relevance of the interfirm network to future policy-making on innovation.

The interfirm network as a level of analysis and policy-making has been neglected whilst attention was focused on the macro level of total economy in economics, and the micro level of the firm in macro organization behaviour. Neo-classical economic theories treated the suppliers as homogeneous from the viewpoint of the firm. Therefore, theoretically, there was little need or economic motive for users to establish long-term linkages to their suppliers and customers. The reasons for neglect in macro organization behaviour arise from an exaggerated concern with 'structural fit' rather than with structuration and innovation. In macro organization behaviour there was the neglect of the dynamic relationship between markets, 'hierarchies', and clans as alternative mechanisms for co-ordinating the deployment and uses of scarce resources. Macro organization behaviour now pays more attention to the role of the market as a co-ordinating mechanism and to the role of subcontracting in providing focal organizations with flexibility, and with a protection against risk arising from fluctuations in demand, and from the risks of investment in embodied capital. The restructuring of economies has disrupted many 'hidden' networks between organizations and to institutions which had been 'silently' sedimented in the past. New technologies are heavily implicated in the transitions of relationships within existing networks. Information technologies have provided both cheaper means of processing routine data with great accuracy and increased the interactive availability of information amongst the members of a trading network. It is now possible for small suppliers to a large-scale subassembler to obtain estimates of future requirements which can be utilized to engage in forward planning. The new technologies can also provide the data bases for subcontracting and for issuing contracts in situations where the contingencies are not too awkward to detect and enforce. For all these reasons, and others, the role of interfirm networks has become an area of enquiry.

In this chapter almost all the references to networks refers to the level of organizations. The focusing of interest on the organizational level provides an important distinction with the enormous literature on individual networks (see Rogers 1987). In fact, the concept of organizational networks and role-sets was introduced by Blau and Scott (1961) and Evan (1966), but largely neglected. Our concern is primarily with the interorganizational level, except for examining the role of individuals in linking amongst organizations and institutions as in the studies of the Stockholm School referred to later in the chapter.

The restructuring of international economies is currently accompanied by the weakening of many interorganizational linkages whose texture and significance for innovation and survival was not fully appreciated. In particular, many vertically-connected firms in established sectors have sought to offload those parts of their operations which are most risky and/or difficult to transform. Also, many sectors contained firms which were implicitly bonded – albeit loosely – in common forms of lobbying, of regulating activity, and of cumulating their learning and recipe knowledge. Over time, sectors became implicitly structured (Whipp and Clark 1986: chapter 2). Some sectors came to occupy key roles relative to their surrounding network. In the engineering industries, for example, the capital goods firms, especially machine tools, implicitly occupied the role of pooling and circulating learning both embodied and unembodied (Rosenberg 1976). Porter (1983b) rightly observes that the dependency on diverse external resources is the life-blood of any enterprise – much more so than the 'structural fit' so admired by orthodox organization design.

Four themes are utilized in this chapter. First, the overall division of labour which emerges between organizations and which often couples organizations as both competitors and collaborators. These relations emerge slowly, often over several decades (Clark and Starkey 1988: chapter 6). The division of labour between organizations is examined later. Second, there is a search for new concepts to describe new phenomena (for example, filière, chains, poles). In a period of economic restructuring, the existing linkages, which had been taken for granted, are often broken apart so that sectors and chains of linked organizations are disassembled. In parallel, though not necessarily in the same location, new sets of linkages are in the process of being formed (see Miles and Snow 1986) often through the role of brokers. That phenomena was outside the analytic domain of macro organization behaviour. However, once the phenomena was recognized there was a search for concepts which directed attention to the organizational network. Organizations rely on many connections and channels, all of which represent a considerable investment of corporate energy and finance. One concept which enjoyed a vogue in the 1980s was 'filière': a multilevel and multifaceted network of channels with clusters and chains running from downstream activities upward to the final customer. Unravelling the key features of the innovation networks – their density and composition into clusters and chains – is the task of this section. Particular attention is given to illustrations drawn from Canada and the UK. These cases illustrate the concepts and also indicate considerable variations in the degree to which some organizations are agentic in constituting the long chains in which they are co-authors of innovative activity. Third, is the theme of interfirm networks as investments and power systems whose hierarchizing can be highly consequential for transitions. Fourth, because this chapter bridges between the diffusion of innovations and their generation and/or ingestation, it is appropriate to consider the interactions between suppliers and users in terms of the themes of the degree of activity of the consumer/user in shaping the eventual form of the innovation. This theme originated with von Hippel's (1982b) notion of the consumer active paradigm, but this perspective is still very heavily oriented to the supplier. The potential for the user to exploit some of the investment costs of innovation is examined by Foxall (1987). This particular theme clearly exemplifies the close interdependence between diffusion and the capability for innovation-design as presented in chapters 9 and 10.

Division of expertises

Guiding themes

A very high proportion of the value added to products and services during their route towards the consumer occurs between enterprises rather than in the final act of consumer purchasing and consumption. Yet the understanding of the dynamics of these interactions has been less salient than the analysis of consumer behaviour. That situation is changing. An important analytic route into the examination of interfirm networks is to examine their relationships as part of a division of expertises within the political economy perspective of macro marketing (Arndt 1983). Two themes may be identified:

1. The long-term dynamics of the division of expertises amongst firms and in the role of the suppliers of expertise (for example, education, professional associations) and of the corporate customers or clients along the chain.
2. The extent to which each of three common modes of co-ordination are utilized and in what combinations: by contract, by hierarchy, or by clan relationships. Co-ordination will probably involve non-market mechanisms in varying combinations.

The aim of this section is to introduce the themes, to report current directions, and to indicate their relevance.

Long-term division

The long-term division of expertises between corporations has not been entirely neglected. Rosenberg's (1976, 1982) examination of inter-dependencies between enterprises and between sectors of the economy is a seminal piece of economic history, particularly in the analysis of the role of the American machine-tool industry in acting as a cumulator and distributor of learning. There are other studies, particularly of the role of consulting engineers in France and in Canada. The latter illustrate a general principle, namely that particular configurations may play or may not play a central role as the repository of expertise, as disseminators and as authors of new forms. Figure 8.1 schematically displays the different positions of consulting engineers in France and in Canada during the 1970s. It appears that French engineers were active, more influential, and more tightly bonded than their Canadian counterparts. The schema is illustrative and situations can alter, especially when established expertises become redundant. An important area of the division of expertise concerns its historical differentiation and changing composition. We can only make a passing reference to this important feature. Figure 8.2 exemplifies the general perspective which might be taken. In this case the differentiation of French expertise in engineering is displayed as a stylized fact. Similar accounts should be developed for other groupings and for other countries.

Amongst organizations the division of expertise is often obscured from immediate view. Whipp and Clark (1986) examined documents and interviewed external suppliers involved in the provision of a new factory and a new car for the speciality division of Rover between 1976 and 1983, and these revealed many crucial interdependencies of expertise of a varied, diverse nature embracing:

* suppliers of paint plants and assembly plants each of which had to provide basic specifications and expertise;
* suppliers of components (for example, stamped bodies) whose expertise in metallurgy and in the interfaces between metals and paint were highly significant;
* architects and contractors responsible for the new plant and its effluent systems;
* the techno-structure of the parent company which acted as a closed gatekeeper towards certain modes of designing the assembly operations and the job content of employees.

Whipp and Clark contend that a rupture emerged between the existing levels of expertise inside the firm and the capability of those within the firm to meaningfully interpret certain developments in expertise externally located. That example raises the issue of the texture of relationships within the overall division of labour.

The approach of the Stockholm School combines two strands: the revision of neo-classical economic theories of the price mechanism as the best form of co-ordination and careful, insightful studies of the division of expertises amongst Swedish firms and universities. First, the critique of neo-classical theory attacks the notion that a focal firm can assume

Figure 8.1 Carrefour model: France and Canada

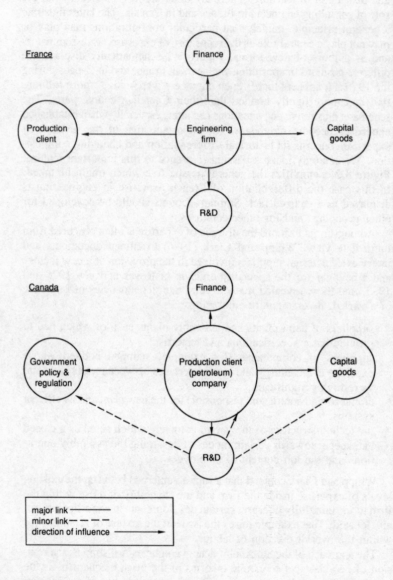

Source: Perrin, cited in DeBresson and Murray (1982).
Note: CEDO = Canadian Economic Development Organization.

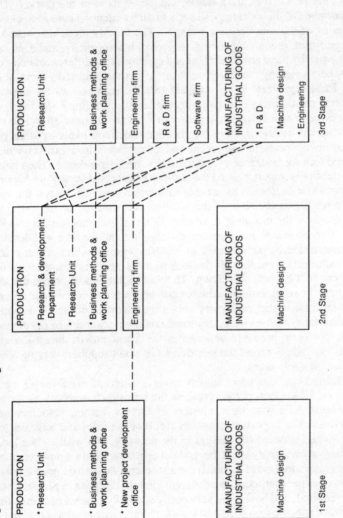

Figure 8.2 The successive stages of the social division of labour

1st Stage

MANUFACTURING OF INDUSTRIAL GOODS

Machine design

PRODUCTION

* Research Unit

* Business methods & work planning office

* New project development office

2nd Stage

MANUFACTURING OF INDUSTRIAL GOODS

Machine design

PRODUCTION

* Research & development Department

* Research Unit

* Business methods & work planning office

Engineering firm

3rd Stage

PRODUCTION

* Research Unit

* Business methods & work planning office

Engineering firm

R & D firm

Software firm

MANUFACTURING OF INDUSTRIAL GOODS

* R & D

* Machine design

* Engineering

Source: Perrin (1977).

that all potential suppliers are essentially homogeneous and therefore offer identical goods/services and exhibit similar attributes: the perfectly homogeneous market. Alderson (1965) replied with the theory of a perfectly heterogeneous market in which each small segment of demand can only be satisfied by a single, unique segment of the market. The importance of this counterposition is to direct attention towards a closer scrutiny of the supplier-user linkages. The Stockholm School assumes a significant, though not perfect, degree of heterogeneity, and contends that typically there are substantial and consequential differences between suppliers whereby the differences can affect the long-term survival of the focal organization. It is argued that the resources to be supplied possess multidimensions which are difficult to determine. So much so that price is an inadequate summary of what the supplier offers. So the user makes decisions about the supplier which are likely to affect the users investments in equipment and other facets in a manner which introduces some degree of specialization and dependence. The choice of supplier is important, and the non-price aspects of supply are important to some degree. In the case of innovation (and efficiency), the user requires a sound grasp of the relative performance of the different attributes of the incoming resources. Detailed understanding is important. Exchange activities between the supplier and the user will develop the initial sketchy picture into a more defined profile which affects costs. The information channel which develops (or is otherwise limited for some reason) will become significant. The Stockholm School argues that the channel is an investment which must be anchored in confidence and mutual trust so that each party abides by the intent of the agreements, so that the expertises necessary to sustain innovation can be developed. This postulate probably understates the latent power dimension and certainly fails to report the worldviews of small suppliers serving some dominant focal users.

Second, the empirical studies cover a range of the Swedish metal sectors. Håkansson (1986) presents the network framework as shown in Figure 8.3a with three clusters of factors: actors, resources and activities. The important questions are: how the roles and activities are distributed between the players in the network, and whether the locus of innovation activity alters the mode of operation from a simple supplier-user pattern to a mode in which the instituters of expertise predominate. How does the user locate the different kinds of complex expertise which are required, and how is the very costly and risky process of cumulating network specific expertise best handled? Particular attention is given to the interfirm interfaces for the situation when a user requires innovations in process technology because this situation typically involves close relationships with the equipment supplier and with other parties such as consultants or specialist university departments. Figure 8.3b depicts

Figure 8.3a Network framework

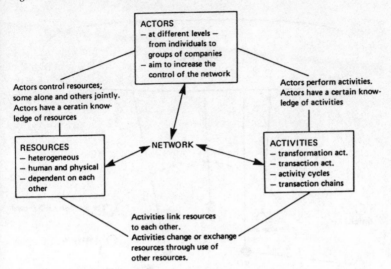

Figure 8.3b Four main categories of process innovation

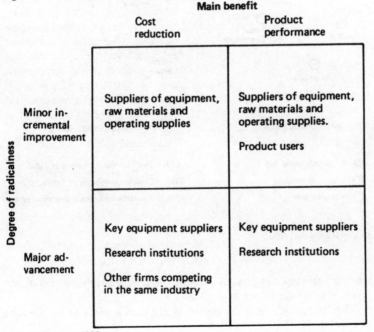

Source: Håkansson (1986).

Innovation-design

Figure 8.4 The Swedish laser-processing network

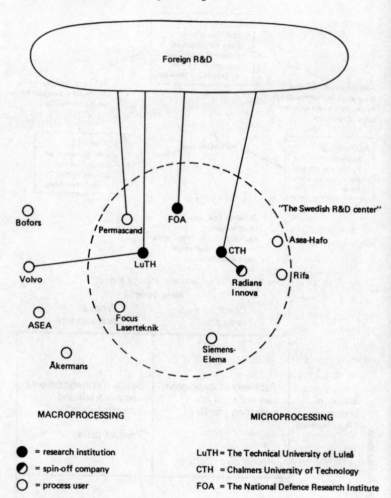

Source: Håkansson (1986).

the four main types of process innovation and suggests the most common types of coupling for each quadrant.

The frameworks and the questions are then applied to a collection of case studies:

166

1. One case study examines and compares how Bofars and Volvo handled the problems of creating a network to handle the development of lasers. As shown in Figure 8.4, each firm relied on foreign research and development, an important fraction of which was mediated through various technological universities. The differences were that Volvo developed a direct, very fruitful relationship with the Lulea Technological University.

2. Another case examined the steel industry and the development of the ASEA-STORA process for the steelworks at Söderfors. In this case there was an extensive potential network of relevant players including: suppliers of equipment, suppliers of raw materials, foreign steel firms, research institutes, and consultants. The Söderfors works developed a key linkage with ASEA, the equipment supplier, as shown in Figure 8.5a,b. At the start the two firms, which were geographically close, were performing parallel development activities which had the potential to be combined. The parties were able to develop a common grammar and syntax for establishing a joint puzzle-solving regime which developed into a trust-based clan relationship.

3. The same equipment suppliers were also involved in another joint development which became known as the ASEA-NYBY process (see Figure 8.5c).

What conclusions are drawn from these and the other case studies? Håkansson's interpretation highlights the high levels of uncertainty which exist in these networks. There are, for instance, vast differences in the time frames of the different players, yet they have to work towards a commonly understood timetable of synchronized activities. The timing of the collaboration is crucial. Typically there is a high degree of trial and error learning which requires perseverance, flexibility, adaptation, and high mutual trust amongst the players. These conclusions confirm the theoretical claim that the examination of the interfirm networks and innovation should start from the postulate that suppliers are heterogeneous and that the price mechanism alone is insufficient as an indicator.

Markets, clans, and hierarchies: combinations

The theory of transaction costs has an important contribution to make to the field of innovation in relation to the issue of interfirm networks. Throughout much of the twentieth century western firms tended to acquire and to vertically integrate the suppliers of expertise not developed inside the enterprise. Some firms prided themselves on their capacity to acquire small firms engaged in prototype innovation and then apply their own expertise in scaling-up operations. A prime British example is the Lucas Company which achieved a central position in the supply of design-led

Figure 8.5a Main actors involved in the development of the ASEA-STORA process

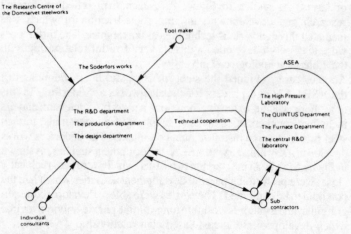

Figure 8.5b Main features of the ASEA-STORA case

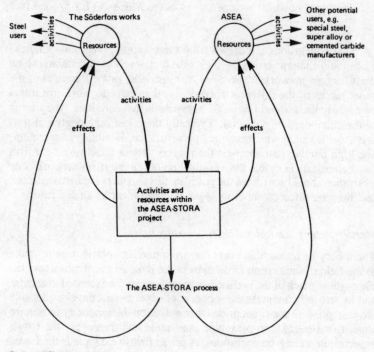

Source: Håkansson (1986).

Figure 8.5c External units involved in NYBY's development of the ASEA-NYBY process

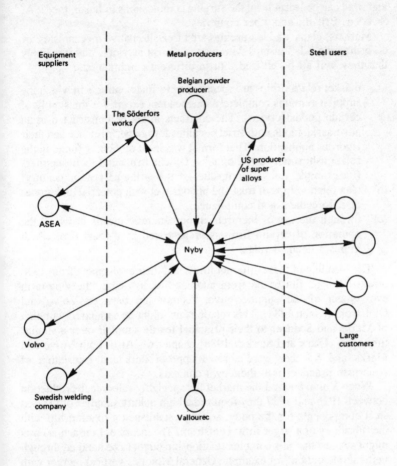

Source: Håkansson (1986).

electricals and electronics for domestic car firms. However, vertical integration often relies on the regular, slow transformation of expertise and so, in periods of 'creative destruction' it can involve considerable costs of investment in outdated expertise. For that reason firms currently attempt to offset their risks from fluctuating demand and from the decay of expertise by relying upon and building up a network of dependent, hierarchical operators. That feature was illustrated in the previous section

169

and is well known in the case of Toyota and its supplier chains (Cusumano 1985). So the issue is when to use particular co-ordinative mechanisms, and what can be learnt from the varying combinations in Japan, the USA, Sweden, Britain, and other countries.

Markets, clans, and hierarchies are three alternative mechanisms for co-ordinating relationships within an interfirm network, and it is likely that they will all be utilized, but in different combinations:

(a) market relationships are specified in formal contracts in which the supplier remains completely independent except for the supply of certain goods/services. The contracts might be constructed in an adversarial mode with strict penalties for nonperformance and their rigorous application. That format was most obviously found in the relationships between the major US car firms and their suppliers (for example, railway companies) in the earlier part of this century;

(b) clan relationships of trust and honour, yet with potential undertones of enforcement and compliance;

(c) through the use of hierarchical/bureaucratic means in which the operation of certain firms is orchestrated by a focal firm which imposes detailed rules.

The blending of these different forms and their evolvement from 1926 onward can be illustrated from another Swedish case: the long-term evolvement of the supplier/buyer relationship between Volvo and Olofstrom (Kinch 1984). This relationship might be compared with that of Marks and Spencer to their principal textile supplier over a similar time span (Clark and Starkey 1988: chapter 6). Also, both Volvo and Marks and Spencer were shrewd appropriators and translators of American practices into their own contexts.

When Volvo entered the market for speciality saloon car production between 1926 and 1927 they required a high quality supplier of pressed steel components for the body, so they established a relationship with the subsidiary of a larger firm: Olofstrom. The theory of transaction costs might suggest that this complex relationship ought to be handled through vertical integration (for example, General Motors's vertical merger with Pressed Steel in 1926), yet from 1926 until 1969 Olofstrom remained independent. Why?

The relationship passed through three periods: 1926–45; 1945–60; after 1960. In the first period the engineers from Olofstrom made careful investigations of American processes in areas like enamelling, and at one stage (in 1937) Volvo sent for ten American experts on coachwork. Volvo undertook the initiative to create the 'spirit of Americanization' at all its suppliers, including Olofstrom. In the relationship between Volvo and Olofstrom there was no written contract, but there was an agreement cemented in personal (that is, clan) relationships which decreed that

Olofstrom would normally be the principal suppliers provided that they equalled American quality standards, delivered on time, and maintained highly competitive prices. In addition they would satisfy Volvo's total needs. This agreement permitted Olofstrom to undertake additional tasks, to invest in new equipment, and to create models of joint learning. Moreover, if Volvo cancelled the contract they promised to pay off the capital investments in equipment and buildings. In 1930 Olofstrom took over the assembly of the bodies. Olofstrom also played a key role in assisting Volvo to solve the problem of an enclosed car – very important in the variable Swedish climate.

The second period was characterized by Volvo's market success with the PV 444 which was selling 25K units annually by 1954 and 80K in 1960. Unwritten agreements were replaced by contracts (for example, 1944, 1950, 1956). Volvo built up an export market in North America and in certain parts of Europe. Growth was particularly dramatic between 1957 and 1960. During this latter period Volvo increased the share of the components produced in its own factories. At Olofstrom these developments meant that the plant increased in size and that Volvo's share of the firms output rose from 50 per cent to 80 per cent. The rapid take-off after 1957 exceeded the forecasts by Volvo and exposed certain limitations to their existing networks of supply, including Olofstrom. However, Volvo increasingly intervened in the operations at Olofstrom to improve quality, quantity, and delivery times. Gradually tensions developed in the relationship which were partly resolved by the creation of interlocking directorships.

In the third period the sales of Volvo rose from 80K units to more than 170K after 1968. Volvo engaged in massive corporate restructuring of its distribution and production system including the opening of a new plant at Torslanda (in 1966) and the formal take-over of Olofstrom and other firms in the mid-1960s. 1961 was the turning point in the relationship, after which its role became less central and parts of its activities were transferred (for example, to Torslanda). By that time Volvo considered that the contract was too favourable to Olofstrom (Kinch 1984: 26). Yet the take-over of Olofstrom at the end of the 1960s was much later than the acquisition of other suppliers. Why?

Kinch reasons that Olofstrom was originally a powerful supplier because it was part of a larger Swedish firm. Moreover, there were highly trustful relationships between the two corporate heads which provided a positive support to the development of expertise in a complementary fashion, particularly in the meeting of American requirements. Subsequently, the proportion of output taken by Volvo increased the dependence between the firms and led later corporate heads to conclude that acquisition was the best route into the future.

The longitudinal case study of Volvo complements the more restricted time frames of the other Swedish studies and reveals the changing combination of the three co-ordination mechanisms. Also, the study of Volvo provides a more incisive exposition of the tensions between alternative forms of co-ordination, and exposes the slightly idealistic vision of trust sketched by Håkansson (1986).

The diversity in the Swedish context is a timely reminder that national profiles tend to be varied, albeit within a 'typical variety', and raises the issue of whether there are cultural predispositions in co-ordination mechanisms. The question should be approached with care because it is evident that the Japanese have modified their domestic practices in overseas settings and that the North Americans and the Europeans are attempting to learn from Japanese practices. However, a brief observation may be offered. The general view is that American firms have made extensive use of formal contracts with the specification of the penalties in the event of contingent clauses being broken. This tendency has been likened to legalism (Crozier 1984) and to the dominance of adversarial relationships over mutual trust. North American firms have also made extensive use of acquisition and that propensity is very clear in the purchase by General Motors and Ford of major software specialists in the 1980s. In comparison it seems that British firms have often made a minimal reliance on strict, tight hierarchical solutions, and have used personal agreements. The Japanese case is varied, yet it is clear that core firms have elected to specialize in the final assembly operations and in the orchestration of complex, hierarchical supply networks which fuse written contracts, the threat of dismissal from the network, and the encouragement of interactive relationships over design. These brief observations point to the requirement for more careful analysis rather than bold conclusions.

Density: societal patterns

The innovation perspective on networks is particularly concerned with the concentration of innovation within and between the interfirm networks. Research has concentrated upon the linkages between firms for major innovations in process technologies at the societal level (for example, DeBresson and Murray 1982; Pavitt, 1984). Such studies are very useful indeed, but they require complementing by investigations at the level of the Swedish examples discussed in the previous section, especially to examine the role of different players along the chains of firms which run from basic raw material suppliers (for example, science-based) to the final distributors (Clark and Starkey 1988: chapter 6). These are examined in the next section. In this section there are two broad issues:

1. The locus of expertise for major, epochal innovation at the societal level.

Table 8.1 Sectoral clusters

		Determinants of technological trajectories				Measured characteristics			
Category of firm (1)	Typical core sectors (2)	Sources of technology (3)	Type of user (4)	Means of appropriation (5)	Technological trajectories (6)	Source of process technology (7)	Relative balance between product and process innovation (8)	Relative size of innovating firms (9)	Intensity and direction of technological diversification (10)
Supplier-dominated	Agriculture Housing Private services Traditional manufacture	Suppliers' research extension services Big users	Price sensitive	Non-technical (e.g. trademarks, marketing, aesthetic design)	Cost-cutting	Suppliers	Process	Small	Low-vertical
Production intensive — Scale-intensive	Bulk materials (steel, glass) Assembly (consumer durables & autos)	PE suppliers R&D	Price sensitive	Process secrecy and know-how Technical lags Patents Dynamic learning economies	Cost-cutting (product design)	In-house: suppliers	Process	Large	High vertical
Production intensive — Specialised suppliers	Machinery Instruments	Design and development users	Performance sensitive	Design know-how Knowledge of users Patents	Product design	In-house: customers	Product	Small	Low concentric
Science-based	Electronics electrical Chemicals	R&D Public science PE	Mixed	R&D know-how Patents Process secrecy and know-how Dynamic learning economies	Mixed	In-house: suppliers	Mixed	Large	Low-vertical High concentric

Source: Pavitt (1984).
Note: PE = Production Engineering Department.

2. Also at the societal level – the linking of different loci of innovations into poles which operate at the sectoral level and in the chains which run from downstream to upstream operations.

Many societal networks lack dense areas, or key linkages are missing, yet the existence of dense linkages should be examined carefully to assess the vintage and relevance of the expertises.

First, the locus of innovation. From a large-scale study of 2,000 British innovations, Pavitt (1984) concludes that a high proportion of the knowledge base which underpinned success was actually located inside the firm in forms which were highly tacit and therefore often difficult to transmit to other firms. Also there were intersectoral differences in the degree to which they generate their own innovations and in their commitment to exporting the innovations to other sectors which rely upon external developments. Four clusters of sectors are identified, as shown in Table 8.1:

1. The supplier-dominated firms (for example, agriculture, traditional manufacturing, textiles) rely upon several external sources, particularly their customers and their science-based suppliers of raw materials and equipment.
2. There are specialist suppliers with their own design and development and knowledge of using patents to develop and protect process designs. They typically have close relationships with their customers and are often quite small, specializing in a particular segment, possibly in products which are later embodied in processes.
3. The production intensive cluster contains firms which are themselves scale intensive in the production of standardized goods in volume and in producing durable goods. Because of their internal exigencies of operation such firms often employ relatively large numbers of experts from different areas, and these technocrats both stimulate epochal innovations and undertake more entrenching-incremental innovation which can be commercially appropriated and marketed to other sectors.
4. The science-based cluster is the prime example of where the locus for new innovations is within the research and development, and design and development of the firms. Often these groups have linkages to specialized research institutes and university departments. Science-led innovations are characteristic of chemicals, pharmaceuticals, and electronic industries.

Table 8.1 shows that the supplier-dominated firms are net receivers, especially from their science-based suppliers and from those scale-intensive firms which also engage in licensing and marketing process innovations. Also, the specialist equipment suppliers are closely related to the science-based cluster and the scale-intensive cluster.

Second, variations in the density of networks at the level of the economy and its major subsystems are considered to be worth mapping in an effort to characterize different kinds of linkages and to examine their possible consequences. This issue has become central in the past decade, particularly in the usage of the notion of filière as the best framework to characterize a connected set of varied channels and concentrations of innovations – innovation poles – surrounding specific sectors of the economy, for example, the electronics sector. Interest in the notion of filière arose in part from the impression that Japanese exporters were targeting their entry into overseas markets to both utilize favourable channels (for example, for distribution), whilst strategically disrupting crucial linkages in the targeted market in order to weaken long-term competition. Studies of the electronic sector in France and of various high-tech French sectors attempting to hold positions in the international division of labour revealed serious weak linkages in the specific filières. A recent overview of the British electronics industry provided a similar profile of commercially successful firms which were only medium sized by global standards and which were disconnected rather than part of an integrated, dense clustering of complementary networks.

The variations in the density of the network and absence or presence of particular clusters of interfirm networks and of chains of linkages running from basic activities through to the final customers is important both as a context for the firm and as a matter of concern for government policy-makers. The directions for further relevant policy-oriented research are evident from the analyses of Canadian innovations viewed in terms of a multiplicity of chains (DeBresson and Murray 1982). The linkages in the Canadian innovative system between 1945 and 1978 were constructed to locate the breeding grounds of major innovation. Canada has less cross-linkages than in the British case (see Figure 8.6) and certain core linkages are absent. There are few nodal points of innovative activity apart from mining, pulp and paper, iron and steel and chemicals, because many sectors are self-supplying enclaves (for example, pulp and paper). There are few dense clusters, particularly by comparison with the UK. Also there is a missing capital goods complex – compared with Britain – and the Canadian equivalent does not play the role of memorizing, cumulating, and redistributing expertise through the economy, and even the consulting agencies for engineering seem to play a restricted role. The trend over three decades is for the few strong nodes to be supplying a decreasing proportion of the innovative effort in Canada. For example, transport, electronics, and electricals are no longer heavy users of Canadian innovations. The Canadian studies illustrate the variations in the density of innovative efforts on a macro-societal level and reveal the absence of key breeding grounds for innovation and the tendency for Canada's best work to be held in national enclaves. Chemicals, for

175

Innovation-design

Figure 8.6 Equipment industries: UK and Canada (circa 1945–75)

UNITED KINGDOM

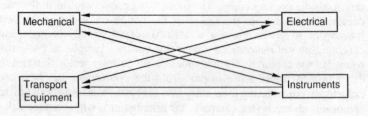

CANADA

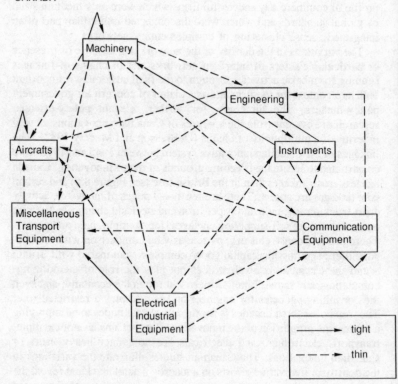

Source: DeBresson and Murray (1982).

example, is a closed-loop clustering. It is the power generation industry which provides the best linkages across sectors.

These examples of the societal level of innovation generation and design require complementing by attention to the constitution of networks as chains within the societal level. This feature is addressed in the next section.

Investments and transitions: constituting chains

The establishment of interfirm linkages is a costly activity requiring the investment in developing channels (for example, around Volvo), and in creating the organizational and collective languages for the regulation and editing of activities. Networks as investments have received slight attention in macro organization behaviour, but there are promising developments in the economic sociology of the French School of Thevenot and Boltanski (1988). Their approach combines elements of the transaction costs and human capital theories, and extends the analysis into the historical development of the City as a location in which differences in wealth could be accommodated within a consensually regulated context. In their empirical studies they have examined the different ways in which these investments are developed by small and large capitals, for example, in studies of the French camembert industry. This approach suggests that constituting chains of linkages is a general requirement for all organizations. However, the notion of chain can be applied at a slightly higher level of analysis which is particularly revealing.

The notion of chains of interfirm linkages is important. It seems obvious that certain firms come to occupy a pivotal role in being the locus of initiatives for generating innovations which become diffused to their suppliers, and for reinventing generic innovations to specify the detailed contingent features of particular contexts. These central firms become nodes for various chains to reach backwards and forwards down the chains of supply and delivery. This tendency has been heightened by the recognition that innovation-design is a specialized corporate activity which may or may not be tightly coupled with other major activities. Miles and Snow (1986) suggest that design is one of four major activity zones with production, supply and distribution. These zones have become loosened and specialized during the economic restructuring of the past two decades, so that brokers play a key linking role. In practice these four activities are probably being recombined in different formats, particularly in relation to the role of finance capital in acquiring and restructuring enterprises. We will return to this point shortly.

It is useful to illustrate the importance of design activity, especially design orchestration of a chain. The role of Marks and Spencer in the

UK since the mid-1920s illustrates the chain. Design is a process which links customer-wants expressed in buying behaviour, to the immediate suppliers of textiles, household goods, and foods. Figure 8.7 summarizes the situation for a major supplier to Marks and Spencer during the early 1960s when Corah (St Margaret) were the single largest suppliers with 5 per cent of the supply line. If we look at the position of Corah from the perspective of Marks and Spencer, it is evident that the latter have developed a design chain which stretches backward to the interfaces with the suppliers of science-based man-made raw materials (for example, ICI and Courtaulds), as well as to the smaller suppliers of specialist knitting equipment (for example, Bentley Cotton) who were developing more advanced machines. That chain had been constituted through the synoptic agency of key figures in Marks and Spencer who had used various analytic and linking devices (for example, their Science Committee chaired by Lord Weizman) to direct attention, and to create forward channels within which the firm occupied a key interpretative role. Since the 1960s there is a great deal of evidence that suggests that Marks and Spencer's designers were highly agentic in enabling and persuading their suppliers to utilize the new raw materials, the new equipment and the new methods (for example, work study from ICI: see Clark 1987). The chain which emerged involved many players other than Marks and Spencer. The chain became a network of channels for evaluating ongoing innovative activity for its members and it developed an implicitly *meta*-level of strategic innovation which affected the long-term survival of textiles in Britain because it became the agency for the transition from nineteenth-century strengths to twentieth-century requirements. Since the mid-1980s certain strains have arisen in the existing chain from the increased specialization of consumer choices and from rival, life-style suppliers. Consequently, the long established firm of Corah faced partial exit in 1988.

This example of the chain requires further application. For example, it seems highly likely that similar though shorter chains emanate from firms in the British food, drink, and tobacco sectors – once one of Britain's most powerful sectors to their suppliers (for example, Rowntree's and Cadbury's; John Player's). Also there were important chains forward to the retailers. However, the growing power of the final distributors has altered the balance of power for the moment to the large-scale retailers who are potentially able to substitute their brand name for all but the most famous names of their suppliers (for example, Mars, Kit Kat, Toblerone, Lite). The distributors – as indicated earlier – are a very important locus for the blending together of complexes which have immense impacts down the whole chain. They are active final agents in matching the main generic technologies to the actual profile of consumer demand.

Figure 8.7 Distribution designers: the Marks and Spencer chain

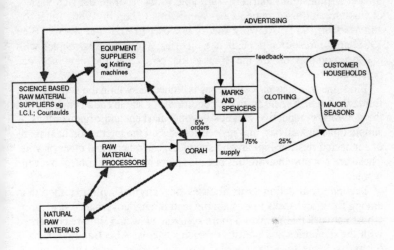

Source: Clark and Starkey (1988).

The retailer-manufacturer example illustrates the case of existing chains which are the outcome of past investments and foresight, but does not highlight how economic restructuring is altering those and providing new opportunities. Clearly existing chains are also political and power systems in which vested interests occupy certain positions. It must therefore be anticipated that new chains are emerging, especially across national boundaries. Chains and the contours both impact the pathway of incoming clusters of innovations, and also are shaped by the inter-action between them and their pre-existing expertises.

Networks

The theory of transaction costs has had an important impact on the current attempt to interpret and prescribe for relationships between organiza-tions, yet its underlying theorizing is static, legalistic, and too systemic. Transaction costs theory tends to fit the situation of just-in-case, and to provide structures of governance suited to the control of plants. Conse-quently the reciprocity of interfirm dynamics associated with just-in-time cannot be readily explained. Yet, just-in-time is becoming a more signifi-cant template than is just-in-case. An alternative theorizing of networks is emerging which offers a more robust connection between the economic and the sociological. Network perspectives – which are not an euphemism for an 'administered market' – are able to incorporate the rich variety of institutional arrangements between firms (Thorelli 1986; Johanson and Mattson 1987). Network theory is closer to exchange theory (Blau 1968) and to resource-dependence theories. We close this chapter with a short comparison of the transaction costs and the new network perspective.

First, the transaction costs theory is embedded within the neo-classical framework which focuses on the conditions for stable equilibrium, whilst the network perspective extends beyond markets and organizations as simple units of analysis, and instead focuses on the interaction in systems of connected relationships among suppliers, customers, and other players. These are not pure hierarchies because parts have some control over the whole.

Second, transaction costs focuses on activities within rather than among firms and seeks to explain the institutional governance structures, whilst network theory takes a more systemic view and is more concerned with the dynamics and politics of such systems. Also transaction costs limits its focus to aggregates of specific transaction relations and is concerned with the examination of the boundaries of markets and organizations. Network perspectives are concerned with the systems of interacting relationships – multiple pairings – in which changes in one set will probably have ramifications throughout the network.

Third, although there are some conceptual similarities, the notion of opportunism, in transaction costs is replaced by the notion of trust as the outcome of the long-term investment by sets of firms in developing interdependency. Network theory gives the formation of knowledge and expertise precedence over notions of bounded rationality. In transaction costs theory a high degree of 'asset specificty' leads to internalization and vertical integration. In the network approach, however, it is itself an outcome of network interactions: a result of the strategic consolidation of long-term relations based on exchange and adaption, and expressed through a 'mutual orientation'. In network theory there is a preference for 'mutual orientation' as the best descriptor of successful interdependencies. Network theory directs attention to the building of lasting, stable relationships between firms based on a highish degree of reciprocal mutuality in the exchange of knowledge and production. Transaction costs theory considers that such mutuality can only exist within hierarchies (for example, bilateral governance).

Fourth, describing the network systems which exist in Japan through transaction costs would omit key socio-economic features which are central to understanding its totality, and to any consideration of extracting specific innovation configurations and transferring these to other countries. There may be two key levels: at the level of the Zaibatsu and at the level of interfirm networks in which just-in-time is involved. Although the earlier form of the Zaibatsu corresponds to a 'clan', its more recent formation corresponds to establishment and consolidation of common interests rather more than trust and honour. Within the modern trade-oriented entities the banking linkages are important (Ruel 1987). Within this level of networks the interlocks from the financial institutions to the manufacturing units do not consider immediate profit as the prime basis for future investment decisions or for strategic planning and investment in research and development. These networks have played an important role in Japanese success and that role explains the slow diffusion to the west of specific items (for example, total quality control) which have been extracted. The other level of the local agglomeration of interacting plants is a precondition for the just-in-time form of organization. Just-in-time is part of a total organizing process which embraces the labour process, the buyer-seller relationship, and the wider context. Suppliers are drawn into just-in-time and to its close relationships which involve the supplier in developing a design competence (cf. Britain). The relationship is continuous, interactive, and reciprocal because the requirement for small, frequent deliveries requires many forms of adjustment throughout the chain to minimize imbalances.

The point is that the network perspective can both illuminate what is essential to an understanding of the new forms of interfirm relationship, whereas the transaction costs perspective, whilst containing some useful elements, is too limited.

Chapter nine

Structural repertoire and corporate expertise

Introduction

We argued in Part I that the abstract, narrow prescriptive frameworks of the orthodox organization design approach failed to address the practical problems of application. Dow (1988) summarizes recent attempts to develop an alternative to the orthodoxy. This chapter presents the perspective of structural repertoires and corporate expertise which provides the analytic framework for examining innovation-design as an integral feature of a corporate multiculture in chapter 10. This perspective is the cutting edge of our approach and whilst building upon the original intents of Burns and Stalker (1961) applies two core features which were undeveloped: the dynamics of structuration (Giddens 1979) and the significance of corporate knowledges (at all levels) which have been defined as corporate expertise. The pathways of innovation in an enterprise are enabled and constrained by the puzzle-solving regimes and scripts embedded in the structural repertoire.

First, the dynamics of structuration are examined through the theory of structural repertoires (Clark 1975, 1976, 1985). The approach differentiates organizations by their capabilities and by the temporal dynamics in which they are actually able to operate. The key distinction between recursiveness in the structural repertoire and transitions of the structural repertoire underpins the perspective. Structural repertoires provide a means of examining organizations as differentiated rather than as homogeneous entities. Differentiation has previously been pursued through the threefold distinction between marketing, production, and research and development (for example, Lawrence and Lorsch 1967). Also, in studies of innovation previous attention was heavily concentrated upon the linkages between basic invention and the role of research and development in initial commercial innovation: the prototype innovation. That focus was particularly evident in American and in British research where the role of research and development was defined as making the bridge between basic science and commercial innovation.

The subsequent states involve a role for research and development, yet that role is directed towards transposing the prototype in new settings. Attention has to be given to design and development as a crucial thrust in corporate viability. That means examining the puzzle-solving regimes which handle the dynamics of the economies of scale and scope.

Second, the expertise-knowledge dimension is addressed by combining the approach to puzzle-solving regimes of Nelson and Winter (1982) with the burgeoning literature on scripts, thinking organizations, multicultures. Our approach combines the already recognized level of the organization with the neglected analysis of the societal context in which knowledge systems are developed (Hoskin and Macve 1988). This chapter pursues that direction with a combination of the following three themes:

1. The analysis of organizations as scripts; the uncovering of how the scripts are constructed and who has the capability to influence their authoring. Particular attention is given to historic and disciplinary framework within which managerial 'work' is undertaken.
2. The notion of puzzle-solving regimes (Nelson and Winter 1977) and corporate thinking practices (for example, Weick 1979). This includes the roles of specific groups like engineers in cumulating expertise and editing its core themes.
3. The choreographing of meanings by top executives to create corporate ideologies which sanction and legitimate particular approaches to exnovation and to innovation.

These themes are woven into the fabric of the perspective on structural repertoires and then utilized to develop the analysis of strategic innovation.

Structural repertoires

Repertoires: scope and richness

The notion of structural repertoires was introduced in chapter 2 with the example from social anthropology of the repertoire of structural poses for a North American tribal society – the Cherokee (Gearing 1958). How can these rudimentary notions be developed into a useful theory of innovation? What are the structural repertoires of organisations?

It is useful to commence by considering the notion of repertoire as:

(a) a collection of actions only some of which can be performed at any one time;
(b) a coupling between a situation and an action from the repertoire. Some situations may have alternative actions available;

(c)   containing some actions in the repertoire capable of activation in anticipation of a future situation to which they are normally coupled;

(d)   containing some actions which are used very infrequently and whose dormancy may lead to their eventual loss from the repertoire;

(e)   one which may be constructed deliberately as in American football teams, or may primarily evolve over time as in the example of the Cherokee;

(f)   varying in their capablity for being edited through deliberate reflection;

(g)   varying in their scope and depth of coverage.

Probably the sharpest, most highly articulated examples of the use of repertoires are to be found with the notion of 'plays'. For example: theatre plays and the plays around which certain American sports are heavily organized (for example, American football, basketball, baseball).

Few organizational studies have applied the perspective of structural repertoire. The notion of repertoire is used in March and Simon at two levels: to refer to individual cognitions and behaviour; to refer to an organization as possessing a 'mosaic of performance programmes'. Their metaphor is with computing and that implies that changes in the repertoire can be accomplished through design and without political processes. The problematics of accomplishing the repertoire are more incisively portrayed in Weick's (1979) perspective on organizing which depicts the organization as a largely hierarchically arranged collection of individuals whose major investment is in interlocked behaviours which can be activated. All members of the organization are involved in usage of enacted definitions of external situations and the process of applying meaning to the equivocality in the external situations is usually problematic.

In order to develop the application of structural repertoires to organizations it is necessary to outline the kinds of structural pose which might be in the repertoire. There are two dimensions: the scope of different categories of organizational repertoire and the richness of the repertoire within and across the categories. The scope of the repertoire can be simply defined with respect to three major categories of organizational situation: the basic operating cycle, the strategic innovation cycle, and special irregular events. These are amplified:

(a) Basic operating cycle

All organizations have operational level units which operate within particular time frames. These units are as diverse as the hospital theatre which is run around the stable temporal unit of the watch so that the full twenty-four-hour cycle is covered, to units like supermarkets which operate on a weekly cycle to educational units which use terms and years . . . and so on. The Cherokee unit was depicted as the year in chapter 2 (see Figure 2.2). Operating units tend to acquire one or more structural poses (Clark 1985). Operating units also possess their characteristic forms of knowledge, and in addition the co-ordination of multi-unit enterprises requires the development of specialized knowledges in co-ordination: the visible hand (Chandler 1977). Co-ordination of the units includes their routine maintenance and that part of the techno-structure dedicated to those sorts of activity.

(b) Strategic innovation cycle

Some organizations develop structural poses to handle the introduction of different forms of operating and which (to varying degrees) orient the adaptive capability towards future problems – the innovational and design structural poses. The structural repertoire for coping with the innovation and design cycle contains a form of knowledge and expertise on which the survival capacity of modern enterprises is grounded. The repertoire of an organization may or may not contain specialized, differentiated poses for handling innovation and design. Medium- and large-scale enterprises normally possess at least one pose.

(c) Specials

Some organizations also develop specialist poses for coping with infre-quently occurring sets of events, for example, to cover emergencies. These are to be found in military organizations (Etzioni 1961: part 4). Many organ-izations do not possess a pose to cope with safety in their repertoire though they claim that their documentary and formalized systems refer to safety. For example, the public enquiry (in 1988) into the fire in London at Kings Cross underground station in which thirty persons died in the previous year indi-cates that the behaviour and cognitions of the staff were not organized into a structural pose. However, it seems that there was a pose to be activated at the Gare de Lyon, Paris (July 1988), when the driver of a train within two minutes of a crashing was enabled to telephone ahead.

Repertoires can be assessed in terms of their performance in use and the measures taken to retain the potential of dormant poses for activa-tion (for example, by rehearsal).

The richness of the repertoire may be judged in terms of the number of poses which cover each of the above three areas and also in the poten-tial for any pose to be used creatively as a framework from which elements may be extracted.

Enactment and activation

Given the repertoire how is it activated? A connection has to be introduced here between external events and the activation of the repertoire. A significant part of the repertoire of structural poses which accretes in corporations is intended to handle events which originate outside the organization and which have direct and indirect impacts on the organization. A key linkage between the external environment and the organization occurs through the flows which make up resource dependency. The issues are twofold: how is the external environment cognitized? and to what degree can the consequences of external events be simulated within organizations? These two issues shape the role of activation.

External events unfold in both patterns which we recognize (albeit much later), and also (we believe) with many chance, irregular events. Evolutionary theories including those of the population ecology perspective emphasize that there are selection criteria embedded in the processes of unfolding, and these criteria decide the extent of key resources which an organization can acquire. The external events cover many dimensions: geophysical disruptions, regular climatic variations, political crises, technological and scientific breakthroughs, and economic pulsations. Within most organizations there are members of the top level élites who seek to interpret those events by constituting cause and effect maps of the external environnment (Clark 1972a; Weick 1979). That process has been referred to as enactment: the superimposing on the external world of simplified, multistate cognitive models and their usage to determine adaptive actions.

Enactment is a socially constructed activity for both players inside organizations and for academics using the 'detour of reflection' to develop *ex ante* explanations. Enactment of the environment can never be complete and must always be selective – even for academic analysts using retrospective reconstruction! Expertise in enactment may contain very sophisticated constructs which are bonded in flexible, yet powerful analyses. Certainly that is what scenario writing attempts to achieve. However, the enactment process is incomplete, and that feature may be heightened in periods of fast external alteration in the events. Yet, it may be reasoned that some organizations contain connected collections of managers in managerial role sets who are both capable in detecting new selection criteria and in achieving the political force to impose their interpretation. That process has been described in several case studies involving the rise to power of new corporate élites and their overthrow of the establishment: for Players (Clark 1972a), ICI (Pettigrew 1985), Chrysler (Reich and Donahue 1985), Ford (Halberstram 1986), Cadbury's (Smith, Child and Rowlinson 1990). These cases tend to support the contentions of Quinn (1980: 104) with respect to logistical integralism. Proponents of the hard-line population ecology perspective will object that the above examples are simply instances of fortuitous adaptation.

The linkage between enactment of the structural repertoire requires a process of activation (Clark 1976). Activation is a complex combination of political, cognitive, and behavioural facets. The activation of the repertoire needs to be examined at two levels: the difference between the routine tactical activation of easily-performed structural poses covering the level of the unit operating systems on the one hand compared with the activation at a strategic innovation level on the other. The tactical level of the activation of sub-units has been implicitly and briefly covered in the social anthropological studies. Activation of the basic repertoire seems to have been relatively simple and widely understood through the constitution of collectively validated grammar to describe key situational events. Studies of the Eskimo and of African tribal societies report seasonal shifts in hunting and other activities. These were often triggered by signals derived from the climate, the flora, and the fauna. These signals were often very clear – though it should be noted that eighteenth-century accounts of British agricultural life (for example, harvesting) show that the signals can be extremely difficult to interpret (for example, see *The Mayor of Casterbridge* by Thomas Hardy). However, according to studies of industrial and retail organizations the detection of the events which are used as signs for changes in the performance of the repertoire may be very contingent – on customer demand, for example (Clark 1985). The signs may be embedded in complex economic activities.

The activation of the strategic level in corporations, the processes of activation, requires the selection, arranging, and interpretation of complex signs whose salience and ordering might well be a matter of considerable debate and tacit skills. Two examples will illustrate the point. Haley's (1972) novel on top decisions in the Detroit automobile industry narrates the desperate political struggle between the established and rising élites over the signs indicating the existence of a new market segment of young, wealthy, sport-loving, outdoor urbanites, and the possibility of their purchasing the compact car. In contrast, a study of top executives in two subsystems of the same textile firm illustrates the diverse signs which one long-established non-graduate group utilized to correctly anticipate a key shift in the purchasing habits of the British customer whilst their graduate colleagues in the other department failed to notice these signs (Clark and Starkey 1988: chapter 6). The reading of external signs in relation to activation may be referred to as strategic time reckoning (Clark 1985). Strategic time reckoning provides the cognitive apparatus within corporations for adapting the organization at the level of its scripts (see next section).

Activation of the repertoire is problematic. Much more so than the computer analogy of repertoire would suggest. There is extensive politization, though that does not necessarily require top level conflict (Hickson *et al.* 1986).

Innovation-design

Repertoires: recursiveness, oscillations, and transitions

The innovation and structural repertoire perspective deals with the structuration of organizations as a totality combining the surface layers of organizational prescriptions (as Pugh and Hickson 1976) with the deeper layers of politics, cognitions, and behaviours. Temporal dynamics, oscillations, and organizational sequences are quite central (Clark 1985). Recursiveness refers to sequences in which cognitions, behaviours and politics are combined into complexes. Frequently recurring structural sequences are endemic in organizations at all temporal levels: the day, week, month, and so on. These sequences are the building blocks of the structural repertoire. Oscillating recursiveness is extensive. The analytic problem is to examine the patterning of recursiveness and to delimit its forms so that the concept of transition of the repertoire can be used much more precisely than has often been the case.

The building blocks of sequences can be edited by organizational members in two very different ways: by custom, and through reflective agency. Custom applies where the main intention of the players is to reproduce past sequences as fully as possible. That is not necessarily as simple as might be imagined. Even in tribal societies, where custom might be thought to be strongest, there is the problem of collective memory and of political agreement. Studies of the initiation rites of tribal societies (for example, the Masai) demonstrate that reproduction often contains unintended alterations. Also, in certain modern societies there is sometimes an established performance of adding custom to modern activities: the inventing of tradition. Reflective agency implies that the building blocks are frequently submitted to observation and revision. That situation is most visible in American team sports like American football, but it also occurs in corporations. Reflective agency may be undertaken within the techno-structure, particularly in corporations operating on the American template (Clark 1987). Its occurrence and application to the structural repertoire of the operating level units may explain why some units actually do achieve incremental innovation – that is, improve the usage of their same level of inputs. Modifications to existing sequences might lead to the significant transition in their operation, possibly through the emergence of novel dissipative structures.

Recursiveness is extensive, therefore the momentum of the structural repertoire is likely to be a major feature of many situations. That is to say, the tendency to retain major sequences in the repertoire over long periods (Crozier 1964; Miller and Friesen 1982). However, momentum is not the simple equivalent of inertia. Momentum implies elaboration around a given archetype rather than inertia. The distinction can be highly significant. It may be that the British hospital service is more typified by momentum than by inertia (Clark and Starkey 1988). Recursiveness

188

occurs both in the operating level units with their short cycles and also in the much longer cycles of activity in the techno-structures which are concerned with long-term adaptive strategic innovation. In a study designed to demonstrate the existence of these longer term sequences, and to unravel their role, Whipp and Clark (1986: chapters 4 and 5) reconstructed the repertoire of a speciality car firm. Successful firms in all sectors have to install in their repertoires those structural poses which can cope with the sequence of major states (not stages) as summarized earlier in Figure 2.2 and examined more fully in chapter 10.

Given that recursiveness in organizations involves both long-term and short-term sequences, how do we decide when the repertoire has been altered? This is a considerable problem, especially for those studies which attempt to examine change through a narrow time-slice of under five years! The problem is to know whether what one observes is the activation of a previously dormant part of the structural repertoire or represents a novel transition in the repertoire. Transitions involve either or both of alterations in: (a) the matching of the repertoire to external events; (b) the composition of the repertoire.

That definition may well invalidate many previous cases of change (Miller and Friesen 1982; Clark and Starkey 1988). Transitions have been dealt with in a cavalier fashion in Macro Organizational Behaviour (MOB). Much more attention must be given to more precise conceptual and empirical studies.

This section has introduced the notion of structural repertoires. The next section deals with their scripting and that leads into the structural repertoire for strategic innovation.

Firm specific knowledges

'Organisations paint their own scenery, observe it through binoculars, and try to find a path through the landscape' (attributed to Tom Lodhal: cited by Weick 1979: 136).

The binding of the structural repertoire to firm specific knowledges occurs within the authoritative network of structuration (Giddens 1979). The authoritative network is situated within the allocative frameworks of the external economic environment. The reproduction of structuration occurs through rules and resource allocation which are recursively implicated. Moreover, structuration enables and constrains individual agency thereby linking individual actions and organization structure (Weick 1979). Structuration is:

(a) inferred from regularities over time and from the reconstruction of the decision rules which generate the flows of symbols and resources through the firm;

(b) a communication network – a diverse, possibly conflict-ridden collection of members whose partial inclusion in the firm creates a multiplicity of cultures and languages;

(c) a provisional configuration amongst competing strategies for power between individuals and groupings. The multiculture is an integral part of what the organization is, rather than being something an organization has (for example, like a coat);

(d) a collection of decision-making routines, especially those concerned with budgets, which provide the institutional constraints within the firm on the acquisition of information and of resources.

The decision-making will be based on bounded cognitions and a coding system which selects signals and identifies the 'time subscripts' which provide the temporal framework of actions and activities.

Structuration as a notion has come on to the agenda of organization studies (Lord 1988). The key linkages are between the structural repertoire and the cognitive dimension of organizational events. The following related constructs are frequently cited: corporate multicultures, scripts, actors' frames of reference, thinking practices, constructs, expertise and puzzle-solving regimes. The common feature across these constructs is that they are concerned with organizational cognitions. Attention to these constructs has often been fragmentary, partial, and insufficiently concentrated upon the innovation-dilemma. Which is the most fruitful analytic route to a useful treatment of the cognitive dynamics of innovation? We suggest that explicating the notions of 'firm specific knowledges' (Pavitt 1987) and of 'puzzle-solving regimes' (Weick 1979; Nelson and Winter 1982) are the most important priority for developing the theory of structural repertoires.

Firm specific knowledges refers to the kinds of knowledge of innovation and of design and development which are developed within specific firms as multicultures. Such knowledge is highly specific to the firm and to its core activities, especially to the interactions between the product knowledge and the production knowledge. The knowledges are often complex because they involve the eclectic combining of knowledge from different sources and their assimilation is costly to the firm. Much of the knowledge is tacit and represents an in-house investment in certain personal constructs which are deployed to steer the enterprise. Firm specific knowledge is specific and often highly cumulative in certain directions for longish periods – even when the efficacy of those directions has become a matter of doubt. Much of this knowledge is derived from developmental work in the techno-structure and yet will often require the development of a linking expertise with basic research and development in the development of prototypes. Whilst we have emphasized the significance of design and development, it is relevant to highlight that many firms have been unprepared for the growing significance of basic science. For example, the development of food science has significantly altered the position of the food manufacturers relative to the downstream suppliers of raw materials and to the upstream distributors.

Firm specific knowledges play a significant role in the viability of the firm, because they mediate between the dynamics of economic activity in the general population of organizations and its impacts on the renewal of the resources on which the firm depends. Chapter 8 demonstrated how firms develop complex external linkages to obtain the blending of knowledge which they require, and which often leads to the articulation and crystallization of an industry paradigm based on common constructs and on similar rules for their interpretation. Industry paradigms are a significant context for firms, yet the possibility of their overthrow and fracturing is always a potential development either through internal re-paradigming or through new entrants.

Scripts as an analogy represent an important concept for describing the firm specific knowledges. The usage of the analogy of script is meant to highlight the recursiveness in the roles played by both individuals and in their interacting role sets (as Merton 1957). An analogy with a theatrical script is being invoked, and it is recognized that the theatrical script will be interpreted by the theatrical producer and by the actor. For example, during the 1980s the British Royal Shakespeare Company performed some of Shakespeare's plays in a modern setting. However, that usage of script leaves untouched a modern variant on scriptwriting for the contemporary media whereby the final script is the outcome of many earlier, negotiated processes between a variety of vested interests – the writer, the client, the technical staff, producers. The process begins with a 'treatment' which is brief, pregnant with key images, multifaceted and open to different lines of development. A client may commission several treatments. Only much later will a treatment be transformed into a script. The full dialogue is added later. Even then the producer (for example, of a video) will add to, re-sequence or severely cut the original script. In the case of a film the final format will be affected by the cutting and editing and there are many silent authors. Likewise in organizations.

In the study of organizational events there have been three distinct usages of script as represented by Goffman, Schanck, and Barley. Goffman (1974) pioneered the subtle use of the theatrical metaphor in which the notion of scripts refers to a loosely arranged, flexible, open collection of already structured utterances which competent organizational players can invoke and deploy according to their definition of the situation. Scripts in this sense are intended to emphasize that variations are constrained and limited, yet the players may be enabled by the script to choose between certain alternatives and to introduce novelty. Goffman's notion of scripts emphasizes that there is a diversity of narratives which are ongoing and their unfolding can be very like James Joyce's account of 16 June in Dublin in the novel Ulysses (1928). Schanck's (1977) approaches contrast sharply with that of Goffman and takes its notion of script from the perspective of artificial intelligence and cognitive psychology. Script has a monochronic, closed format. Barley (1986) prises open the open/closed contrast in the usage of the script analogy in his examination of the introduction of new medical technology into two unconnected groups of radiologists.

Nelson and Winter (1982: chapter 5) concentrate upon puzzle-solving as an organizational activity which is central to the roles of the techno-structure in strategic innovation. The puzzle-solving regimes associated with the innovation configuration are unlikely to possess the tightly-bonded character of scientific paradigms and exemplars. Therefore Weick's (1979) notion of organizing as puzzle-solving is an appropriate contribution to introduce. Weick's theoretical framework on organizing aims to provide a holistic perspective in which the analyst searches for the multiplicity of causal loops and causal networks around which organizational cognitions and activity circulate. Recursiveness is both chronic and precarious. Without recursiveness organizations do not exist. Collective recursiveness in organizations involves the constituting of behavioural interlocks between players into structural poses. The interlocks between players are the crucial elements rather than the whole players – there is partial inclusion. There is a wide repertoire of interlocks which may be constituted into a structural pose. Their activation requires a cognitive input in which there is a causal cycle which consists of streams of events on the left to the interlocked behaviours on the right through puzzles, interpretation by recipes, and the building of organizational procedures. Weick postulates that players face streams of events which contain considerable randomness and chaos as well as certain arrays of sequences which the players believe possess a patterning. The players cognitize the stream of events to impose generality, reliability, and simplicity. From the streams of events, both external and internal, the players face puzzles which they attempt to interpret by applying organizational recipes derived from past experiences and subject to some degree of modification. These recipes are largely taken for granted, deeply sedimented, almost tacit. The recipes form the grammar of the corporate script. Recipes contain causal loops and they activate causal networks of procedures and interlocked behaviours. The importance of recipes is highlighted by addressing a key problem in orthodox organizational design – the production of blueprints: what recipes will produce a given blueprint and do these exist in the retention system of the organization (Weick 1979; 47)?

The development of a firm specific recipe states how behaviours can be combined into procedures. These recipes may be unrecognized by many players inside an organization. Meta-recipes will define resource dependency and resource allocation. Hall (1984) shows that the demise of the *Saturday Evening Post* could be attributed to the meta-recipe which related increases in the size of the advertising budget through a fixed ratio to increases in the size of the edited copy, so that the weight and size of the journal grew beyond the economies of production and distribution. Such meta-recipes are increasingly common, not only in public sector organizations (for example, health, education) but also in the private sector where they constitute the 'hard' framework around which much politicking occurs. In addition, there will be a skein of recipes which permeate all aspects of organizing and many will be tacit. The firm specific grammar implies the existence of a consensually agreed grammar, but Weick argues that minority rule is pervasive, often through the placing of contentious decisions into zones of indifference. Typically consensus is within an asymmetrical power situation (Giddens 1979) and may involve a high degree of hegemonic influence (Burawoy 1984). Examining the consensually accepted grammar requires an ability to reconstruct the many feedback loops and linkages on which organizing is erected. It will be important to examine 'plans' and to theorize these as symbolic adverts for the future.

The recipes mediate between the translation of equivocal streams of events, and the constitution of interlocked behaviours into procedures. The interlocked behaviours which connect individuals and constitute a major part of structural pose are assembled into sequences and into interdependent actions through verbal and tacit cognitions. There are many interlocked behaviours in the repertoire and their composition into sequences is mediated through the interpretation which is placed on events, yet events are sometimes puzzling. The puzzles contain varying degrees of equivocality which have to be decoded and encoded through the application of a recipe before procedures and behaviours can be deployed.

Nelson and Winter maintain that established puzzle-solving regimes provide the cognitive boundaries to the degree of equivocality which can be absorbed and fruitfully deployed to improve existing capacities. Moreover, it seems likely that these regimes are closely connected with the career tracks of a particular cohort of individuals (see Burns and Stalker 1961: chapter 6). These cohorts may have been implicitly bonded through particular organizational sagas. For example, the cohorts who came into power in Renault during the 1950s were formed in the resistance and articulated their script for the firm around the 4CV, and its influence remained evident well into the 1950s after which they were challenged by another cohort (Clark and Windebank 1985). Pettigrew

(1985) documents a cohort-script struggle in ICI which preceded the switch of the firm into speciality chemicals.

Politics and power

Organization studies is still deeply shaped in apolitical forms of theorizing in which the political and power dimensions are bracketed. In the perspective of structuration, politics and power are integral and multifaceted. Conflicts will occur down many lines of cleavage between vested interests and these will only occasionally become manifest in overt clashes. We have illustrated the dynamics of struggle between capital, management, and labour with reference to the design process (see Figure 2.3). In that case study there were many dimensions to the conflict, especially when the management lost its capacity to keep certain issues off the political agenda.

Chapter ten

Innovation-design as corporate culture

Introduction

The combining of innovation and design into the innovation-design perspective has been constructed to counter the dual dangers of the pro-innovation bias and the pro-efficiency bias. This chapter shows that, within organizations, the processes of innovation require the combining and blending of two sets of activities, whose distinctive features are often unrecognized by practising managers, and whose complementarities are unrecognized by academics. The two activities are:

1. The strategic design cycles through which products and services move periodically as new models replace old models (innovation-design).
2. The ingesting of externally developed innovations (innovation-diffusion).

The orderly separation of these two activities for analysis contrasts with their intermingling in the actual context of organizations. Most organizations face a plurality of externally-generated, divergent innovations, some with complex configurations. Consequently, the periodic cycles of innovation-design are usually impacted by the ongoing diffusion of new materials, techniques, and practices. Moreover, there may be a problem of exnovating existing practices.

In the changed market context the pace of model replacement has increased. In clothing, for example, the slow pace of the two-season year began to collapse in the 1960s, and today there is a four-season year for many consumers. Also, the wearer has begun to segment the day into different activities each typified by different styles of clothing. Discovering what the consumer wants is more difficult – even for the consumer. Uncovering these wants has a strong trial and error element in which the actual patterns of consumer purchases are used to trigger reordering. The suppliers offer a wide scope of garments which can be ordered in small batches. The new market situation changes the economies of scale and scope, and transforms the relationships between distributers and their suppliers. The new market context has been stimulated and created by the diverse arrays of innovation in raw materials, in the capabilities of equipment, in the logistics of economic activity, and in new forms of organization structure. Adapting to the demands of the new market situation requires a combination of commitment and flexibility: the corporate culture of innovation-design.

The aim of this chapter is to also draw together the diffusion and design themes. With innovation-design we focus upon that part of the structural repertoire which contains the design cycles and which has to interface with ongoing operating systems. In the area of innovation-diffusion we present a multistate decision episode framework which incorporates the influence of external factors on the decision to adopt and of mainly internal factors on the attempt to implement. We argue that organizations need to develop specialist skills in innovation appropriation which include the capacity to unbundle innovations.

Strategic design cycles

Design as total process

Kantrow (1980) contends that the area of strategy must be reintegrated with the topic of innovation so that strategy, innovation, and organization are treated holistically as a total process. That process is increasingly referred to as design, and it is recognized that designing is a cross-functional activity which links the external market to the internal capabilities of the enterprise. Design is central to products (for example, personal computers) and services (for example, 'Club Med'); to production processes; to channels of distribution; to supply chains; to image, and so on. Design cannot be equated with aesthetics even though such considerations cannot be excluded. Nor can design be simply thought of as organization design.

Design should be envisaged as a constellation of processes involving many players and directed towards internalizing customer requirements and interests. In most organizations particular moments arise when there are opportunities for existing designs to be strengthened and made more robust or for new paradigms of design to be introduced. The danger is that these opportunities are lost in the pursuit of fragmented objectives: lean designing. That situation arose in the British automobile industry with the Mini and subsequent models. The objective should be the multifacet creation of a grammar which can be stretched and reconfigured to incorporate ongoing opportunities and developments. That does not mean the incorporation of massive diversity. Many successful designs of products and of processes are not at the leading edge of being the most innovative, rather they are using the learning experience of the early adopters. Their incorporations concentrate upon the core attributes of generic and epochal innovations.

We shall consider design to apply to four areas considered jointly: the product (or service), the means through which it is produced (including the supply chain), the means of distribution, and the organization processes. The crucial objective is to discover the competitive edges of the organization (Porter 1985). It is even likely that a good product design may be coupled with poor organization design. Design became established as a part of the structural repertoire in the earlier part of the century for a small number of very large-scale organizations. Now design is a general requirement.

Design states

There is agreement that designing as an activity should and does involve an array of states which ought to possess a certain sequential cumulation whilst also displaying considerable iterations, abortions, culs-de-sac, and chaos. There are many multistage models for the prescribing of design, and many organizations have developed elaborate, equipment-based procedures for designing. We have chosen a four-state framework consisting of vision, translation, commissioning, and operating. The framework was illustrated earlier in Figure 2.3 with the cycle of the design for a new car. Visioning is not confined to the design of the product. It also involves choices about the means for production and distribution and the forms of appropriate organization (cf. Miles and Snow 1978: 24).

First, visioning centres on the strategic choices of the product or service and market domains: their scope, diversity, and geographical marketing (for example, Europe versus North America). For example, in the past two decades three leading British firms have attempted to reshape their product markets: ICI, Cadbury, and Rover. ICI, the major chemical firm, has shifted from the production and distribution of general chemicals sold in domestic, Commonwealth and European markets, to speciality chemicals sold in North America and continental Europe. Cadbury-Schweppes have narrowed their wide, seasonally-variable product range into a narrower range of products which can be distributed throughout the year to different regions through improved manufacture and temperature-controlled distribution. Rover intended to design a high volume speciality saloon car (see Figure 2.3 and chapter 5) in direct competition with leading European models: BMW, Mercedes, Saab, Volvo. Until that time no British-owned car firm had succeeded in making a significant penetration of the European market. The car design was certainly oriented to Europe and incorporated styling features which echoed the larger, sleek-lined Citroens, yet British car firms lacked the marketing infrastructure to thoroughly encode the relevant tastes and to distribute the saloons. With the advantage of hindsight it might have been better to have aimed for the American market and used the Jaguar marque which at that time was in the product range. Visioning the market will involve a choice between local and global markets (Ghoshal and Bartlett 1987).

The choice of the product and its markets is the key strategic decision because they shape the scope and scale dynamics: customized, speciality, or general. Also the location in which products are distributed is a major transaction with the pools of resources on which the organization depends. In chapter 7 we suggested that markets have their distinctive temporal features (for example, pace of saturation), and that these features have the potential to entrain the organization in ways which may make adaptation between markets a major source of problems.

A further key problem in visioning is the anticipation of those strategic issues where there are discontinuities emerging which will impact the structural repertoire of the organization. For example, in the 1970s the 'big 3' American car firms probably failed to anticipate the threat of foreign imports (Sobel 1984), possibly because the normative frameworks of the car-making community in Detroit was close-minded to the discontinuities (Yates 1983). Discontinuities are not readily detected from those forecasting techniques which simply extrapolate the past. Hence the development of procedures like scenario writing. Visioning is the start of a long-term probing into the future. The length of the time-scale may vary from fifteen years for special military projects, to almost ten years in the case of hospitals, to around four years for a new automobile, to eighteen months for a variety of products and services including new degree courses. Given that this stage of designing can often be very costly, and that its decisions can dedicate a high proportion of future opportunities, then the paradigmatic base is highly important. The importance of this feature is recognized in the current efforts of many enterprises to acquire significant forms of expertise in systemofacture through the acquisition of companies specializing in information and its related technologies. Likewise an increasing number of organizations are seeking to describe their paradigmatic base (for example, Labatt) as one step towards its scrutiny and possible revision. The outcome of visioning is an architecture consisting of an image of the future with its basic grammar largely defined and with the key parameters in place. These may be revised – perhaps considerably. On the other hand, the vision may drive subsequent states (Ghoshal and Bartlett 1987).

Translating is the second state and consists of transformation of the broad parameters of visioning into specific features. This area has been dealt with extensively in the standard texts, yet they have largely neglected the enabling and constraining impacts of the existing structural repertoire (Whipp and Clark 1986: chapters 4 and 5).

Commissioning is a major trial and is often tense, sometimes exciting. It consists of the initial trials with the new product in its new plant and supply lines. Figure 2.3 shows how Rover introduced their new car into its new paint plant and on to the assembly line by running through a single car, then slowly increasing the number to detect and rectify production problems. Many of these might have been anticipated. Commissioning can take a good deal of time, especially where altering innovations have been incorporated. In hospitals, for example, there has been a tendency to transfer activities previously undertaken in the wards to a central treatment area (a new-built environment) where new equipment has been concentrated to increase its utilization and to enable maintenance. Commissioning these areas is an anxious time for the contractors and for the users.

Operating the new systems is the final state. This may be viewed as the end state for strategic innovation, yet there is a great deal of incremental innovation which might be undertaken to improve overall effectiveness.

Robust designing and silent designers

Designing involves the choice of a set of directions whose cumulative impacts can be fateful. A distinction can be drawn between lean and robust designing. Lean designing is when the costs of investment in design cannot be recouped because the initial choices of direction raise marginal costs above marginal revenue. The prime example is the Mini, a popular automobile with many novel features created out of developmental investment by Rover. The Mini sold more than five million over more than a quarter of a century yet barely covered the costs of its complex development and certainly contributed very little to the financial viability of its designers. The example of the Mini shows that too much can be attempted in designing relative to what the market will be willing to pay. The main rival to the Mini was the Ford Prefect, which some will remember as a functional, simple box-like vehicle, with no frills and a very basic, reliable performance.

Rothwell and Gardiner (1983, 1985) maintain that robust design involves an initial stage when considerable diversity is compacted into a new visioning and translation which is then – at a later date – stretched into a new family of models. However, the comparison of the Mini and the Prefect cautions against the necessity to ingest high levels of diversity. Robust designing may involve the ingesting of considerable diversity, yet that is not a necessary condition. The key issue is the balance between the stretchability of innovation configuration (chapter 3), and the degree to which the output matches the resource pools on which the organization depends. The Land Rover, developed in 1947, for example, has been stretched into many uses and a new marque, the Range Rover, yet its initial design blended the military jeep with the existing design capability developed by Rover over the previous fifteen years.

Who does the designing? Considerable attention has been given to notions like product champions, yet most current research in the marketing of industrial products (for example, Bristor and Ryan 1987) and in the running of design activities, emphasizes that there are many silent designers (Dumas 1988). The significance is that innovation and design have to be total processes in which there are many 'authors' of the future.

Ingesting: decision episode framework

Earlier we noted the tendency of innovation diffusion perspectives to limit their analytic focus to the mechanisms which inform the processes by which innovations travel through space and time from invention to adoption (see chapter 6). We also noted some of the limitations of such approaches and, in particular, that the moment of adoption of an externally-generated innovation should be seen as the starting point, as opposed to an end point, for the analysis of the innovation process. The clear requirement here is for a perspective which attempts to bridge the gap between diffusion and organizational studies through an analysis of the implementation or ingestion of innovations. This area has been the subject of much research and theoretical speculation. The contribution which we wish to highlight here is the treatment of innovation in organizations which centres on the collaboration of E.M. Rogers and J.D. Eveland (Rogers and Rogers 1976).

This work on innovation in organization is influenced by the contingency and planned approach of Zaltman and Duncan (1977). Earlier models of organizational innovation as a time-based activity identified two distinct stages and two distinct organizational structures. The stages are 'initiation' and 'implementation'. Initiation refers to the process by which an organization becomes aware of an innovation and decides to adopt it. Implementation refers to the process by which an organization puts the innovation into practice. Such dual structure theories argue that those structural characteristics of an organization that facilitate initiation make it difficult for the organization to implement the innovation. Thus an organization which is characterized by high complexity, low formalization, and low centralization, is more likely to adopt because it will tend to be more open to its environment. The successful implementation of innovations, on the other hand, requires a structure characterized by high centralization, high formalization, and low complexity. Rogers recognizes the value of such observations, but points out that although structural characteristics represent important contextualizing features, such dual structures may be impractical or impossible for most organizations to manage, and so he attempts to develop a more differentiated model of the process of innovation in organizations.

The organizational model of innovation, Figure 10.1, develops the processual understanding of innovation and expands the scope of analysis: now the innovation process is itself contextualized in the broader environment in which the organization is situated. The model has five main elements:

1. Internal and external influences upon adopter units are given a more comprehensive treatment which goes beyond the closed boundaries of individual psychology and community tradition to include a number of factors: knowledge of possible innovations, demands and supports, and resource constraints.
2. The adoption decision is the outcome of an active process of problem definition as opposed to a passive acceptance or rejection of supplier initiated possibilities.
3. Innovation is but one of a number of possible responses by the organization to a confluence of external forces and organizational requirements. Innovation may not always be the optimum course and other responses may in some circumstances be more appropriate.
4. Innovation proceeds through a series of closely-related stages representing increasing commitment by the organization. These stages are neither discrete nor unilinear, but interactive and subject to an array of loopbacks and modifications.
5. The outcomes of innovation are considered in the broader terms of organizational effectiveness.

First, the prior existence or active creation of a stock of knowledge concerning particular innovations and the perception of their appropriateness is seen as an important influence on the likelihood of an organization bringing an innovation across its boundary and incorporating it into its structure. The model marks a crucial shift away from earlier formulations in the recognition of the role of organizational knowledge, not just at the matching stage, but also throughout the whole innovation process. In particular, the conception of an innovation as a static entity which crosses the organizational boundary to be accommodated relatively unchanged into the ongoing practices of the organization, is rejected in favour of a more dynamic understanding. Innovations 'go through extensive revision, essentially amounting to "reinvention" in the process of their adoption and implementation within the organisation' (Rogers and Rogers 1976: 159). Thus organizations often adopt a general concept whose operational meaning gradually unfolds in the process of implementation, as opposed to a specific blueprint for an innovation.

Innovation-design

Figure 10.1 Paradigm of the innovation process in organizations

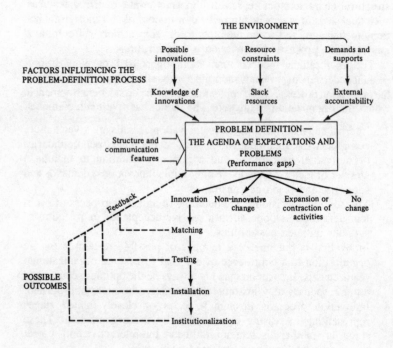

Source: Rogers, E.M. and Rogers, R.K. (1976).

The notion of 'external accountability' refers to the degree to which an organization is dependent on or responsible to its environment which may be defined in relation to various organizational boundaries. It entails the recognition that the innovation process is not a series of events which occur exclusively within the boundaries of a specific organization. The whole process is grounded in, and contextualized by, its links with external networks, pressures, and forces. Thus the 'external accountability' of an organization is an important contextualizing concept. Such dependencies may be defined in terms of the need for funds, specialized personnel, psychological elements (for example, significant reference groups), and operational dependencies. Operational dependencies may be of two kinds which correspond to the nature and scope of legitimate questions to which an organization must respond. Other exernalities include the location of organizations or specific personnel in inter-organizational networks which co-ordinate the flow of knowledge in various forms. For example, the involvement of organizational staff in interorganizational relationships, over time, which exposes them to new ideas, new perspectives, and experience of new approaches and techniques to organizational problem solving. These connections may be expressed in the form of the existence or establishment of collegiate relationships, or they may be hidden elements in the mobility patterns of key personnel over time. External accountability is a crucial element in the generation of shifts in organizational expectations leading to the perception of performance gaps. In general, the greater the number of external accountability relationships to which an organization is a party, the more expectations it will acquire, and the probability of it embarking on a search for solutions. External accountability is also instrumental in the distribution of knowledge and experience across organizations through time.

Whilst the ability of an organization to generate the kinds of awareness necessary to the identification of performance gaps is closely related to the interorganizational context in which it operates, the capacity of an organization to respond to identified problems is also subject to intraorganizational elements, particularly concerning the availability of a potential for mobilizing necessary resources. In relation to this the concept of 'slack resources' (Cyert and March 1963) is recognized by Rogers as an important one, but he is critical of the lack of depth and precision with which it has previously been operationalized. It refers the existence of resources not already committed to other purposes which facilitate the absorption of pressures generated in the adoption and implementation process. All innovations require an investment of resources for their implementation and the prior existence of slack resources is an essential ingredient in the innovation process. In Rogers's formulation the concept is used to include a number of dimensions:

(a) financial slack which includes the amount of long-term commitments, the amount of shift between budget categories over time, the amount of discretionary funds, amount of total budget increase, the amount of difficulty in securing new funds, and so on;
(b) personnel slack which might include the number of temporary or short-term employees, turnover, work loads, and the number of professionals not working full time at their profession;
(c) physical slack which includes unoccupied office slack, accumulated office supplies and equipment, computer down-time, and changes in service-demand patterns. These three categories are intended to be illustrative, as opposed to all-inclusive. Slack may already exist in an organization, or it may be intentionally created.

The amount of organizational slack is closely related to the likelihood of adoption. However, it should be noted that different kinds of pressure or slack may have different influences at different moments in the overall innovation process. It is not the existence of slack in itself that is important, but its combination with other general indices and in the pattern of relationships between various types of slack resources. Thus, a combination of increased physical pressure (for example, increased market demand coupled with financial slack in terms of the availability of funds) may facilitate adoption, whereas physical slack in terms of a temporary reduction in demand level may facilitate implementation as pressure on the performance of the innovation and organizational personnel is reduced at a crucial moment. Financial slack created through the exertion of pressure on personnel levels is likely to facilitate adoption and yet hinder implementation.

Second, the innovation process is itself preceded by a period of problem definition in which the agenda of expectations and problems is set. This represents an important improvement on earlier formulations on two counts. First, the earlier model held individual adopters to be largely passive actors in a process in which diffusion agencies remove obstacles to information flow and thus to adoption. Now potential adopters are conceptualized as active participants who intentionally scan the environment for solutions as opposed to receptors of information diffused by agencies. This period of agenda formation is conceptualized in terms of a dual process of problem definition and the formulation of solutions. Problem definition concerns the process whereby an organization becomes aware of, and defines, specific 'performance gaps' which correspond to an identified discrepancy between an organization's expectations and its actual performance. Such definitions are subject to variable degrees of specificity ranging from loose, general statements of concern, to operationally-precise prescriptions for action. The key feature of a performance gap is that it describes a dissonance between expectations and reality. Thus gaps may be created or removed by changes in either dimension. The perception or lack of perception of performance gaps is mediated by organizational context and environment: expectations may, for instance, change as a result of an external shock, for example, a change in the market which may create new realities, or a change in ownership, or the infusion of new blood which may provide fresh vision and create new expectations. Perception of a performance gap may result from either, or a combination of the two.

Third, it is important to recognize that this stage takes place prior to the process of innovation itself. Problem definition is a precondition for innovation as opposed to a stage in the innovation process proper. The two processes are not necessarily tied in an undirectional way. This is important because it most clearly marks the shift away from a pro-innovation bias: the outcome of this period may be a course other than innovation, for example, non-innovative change, or expansion or contraction of activities. Thus this part of the model represents a contextualizing element for a possible adoption of an innovation.

Fourth, in place of the more usual two broad phases of initiation and implementation, the model consists of a number of discreet states all of which are subjected to pressures both from the environment and as a result of the decisions and actions taken during preceding stages. Four states are articulated: adoption or matching, testing, installation, and institutionalization. Each state engenders a range of issues and problems which need to be negotiated in the process of the selection of specific innovations at both generic and subsystem level. The adoption decision marks the beginning as opposed to the end of the innovation process. The first state involves the matching of an organizational problem (performance gap) with a potential solution in the form of an innovation. This marks a shift from problem definition to innovation proper. The testing stage represents a limited implementation to assess the accuracy of the match and its possible effects. Installation refers to the process of connecting the innovation to the ongoing structure and activities of the organisation. Institutionalisation refers to the process of removing the status of 'innovation' from the new element, thus making it an integral part of the system.

Fifth, innovation itself is displaced as the major research focus in favour of a more sophisticated process-based formulation of innovation:

> Innovation . . . is the process of adopting new ideas. Innovativeness, on the other hand, is a property of the adopting unit . . . In the process terms we use in this chapter, innovativeness is not the ultimate dependent variable, but more of an intervening variable that is predictive of an organisation's effectiveness.
>
> (Rogers and Rogers 1976: 151–3)

Here notions of innovativeness, which characterized earlier formulations, are clearly relegated to a secondary position. Further, the model contains a clear repudiation of the pro-innovation bias through the recognition that decisions concerning innovation are but one option open to an organization at any given time, all of which must be considered in terms beyond themselves: the innovation process can only be considered necessary or successful in relation to broader organizational objectives. The major variable is now seen as organizational effectiveness: that is the degree to which organizational purposes are achieved. There are, of course, difficulties surrounding a clear formulation of such a concept and a system-demand model is put forward which concerns measures of the ability of an organization to mobilize to meet demands in the areas of production, adaptability, and flexibility.

Exnovation and innovation

We commenced in chapter 1 with a short quotation from Schumpeter on the notion of creative destruction, and then in chapter 5 we referred to the occurrence of irregular periods when there is a massive restructuring of capital formation and massive dislocation of existing social institutions. Yet we have also reasoned that social institutions are not purposive and are not readily available to be redesigned: the arguments from design. Instead we have argued that decision-making routines in organizations typically have longish periods of slow, incremental revision, and that these may be decreasingly capable of directing organizational action towards the crucial resources required for renewal. These routines are an integral part of the multicultures and they are entirely politicized. Existing structuration may inhibit innovation. So innovations may occur through the exit of existing organizations and the entry of different modes of organizing in a population ecology effect. However, we have reasoned that some organizations do find the decision-making capabilities to achieve an adaptive fit through innovation-design. Such organizations will – from time to time – face the problem of exnovating existing practices. This final section points to exnovation as a problem requiring analytical and practical attention, but one which is not well understood.

Exnovation by corporate members of existing innovation configurations has been largely ignored (Kimberly 1981). There are two conditions under which the removal of existing practices may be secondary: the existing modes of organizing cannot be adjusted to the new circumstances so the host organization will exit in the near future; or, the existing practices can be left momentarily intact because an overlayer of decision-making can be superimposed. The former situation is important, and the Miller-Friesen (1984) archetypes suggest modes which fall into this category under certain conditions. The latter situation does occur. For example, in the case of the Rover firm in 1932, its subsequent revival left the structural repertoire of the operating level in the plants largely undisturbed, whilst the board differentiated and superimposed a new repertoire of design structures (Whipp and Clark 1986: chapter 3). Likewise, one of our ongoing studies of the introduction of computer-aided production management into the plastics sector contains an interesting example of the superimposition of a corporate design level on top of previously small-scale, quasi-autonomous operating units. However, these two situations still leave the problem that in many situations there are existing, highly-routinized practices.

The problem of the exnovation of existing practices may be illustrated from the study by Yin (1979) of the introduction of new practices into local service organizations such as the police or the fire service. There are some parallels with the Rogers/Eveland schema, but Yin reveals the problem of exnovation more sharply by sketching a stylized life-history for exnovation and for innovation. Figure 10.2 depicts the schematic

Figure 10.2 Exnovation and innovation

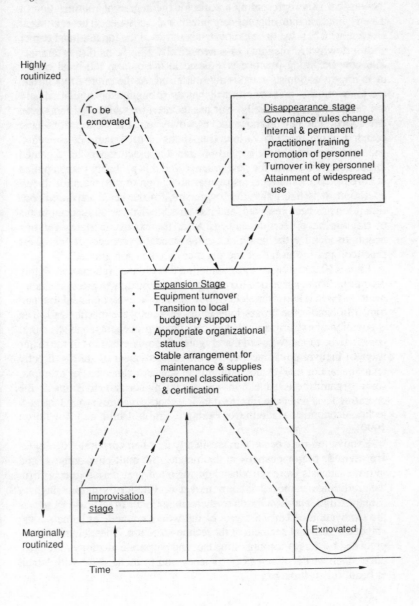

Highly
routinized

To be
exnovated

Disappearance stage
*   Governance rules change
*   Internal & permanent
    practitioner training
*   Promotion of personnel
*   Turnover in key personnel
*   Attainment of widespread
    use

Expansion Stage
*   Equipment turnover
*   Transition to local
    budgetary support
*   Appropriate organizational
    status
*   Stable arrangement for
    maintenance & supplies
*   Personnel classification
    & certification

Improvisation
stage

Exnovated

Marginally
routinized

Time

Source: Yin, R.W. (1979).

presentation with the temporal dimension along the horizontal axis and the vertical axis representing a scale for the degree of routinization of a new practice into ongoing organizational activities. The routinized practice which is due to be exnovated is shown at the top left-hand corner with a downward diagonal to schematically signify its disappearance. The new incoming practice is situated in the bottom left-hand corner in an improvization state and is not routinized. At the improvization state the innovation is just operational, but its strengths and limitations are not yet known experientially. Nor has its interface with subsystems been developed. The new practice moves slowly towards the top right-hand corner passing through various transitions in its 'bundle of elements' and in their uses. At the expansion state the practitioners have gained direct experience with the new practice and it is probably being applied to core practices and receives regular allocation of resources for its full operation. It is likely that the 'ownership' of the new innovation con-figuration has been resolved, and it is possible that various special forms of training have been developed. After the expansion state everyone begins to identify the new practice as standard practice (cf. new best practice) and the usage of the practice is taken for granted.

Figure 10.2 highlights the assumption that old innovations will simply disappear. That expectation may apply very adequately to pieces of equip-ment, yet we all know the story of how artillery crews continued to retain a man to handle the horses long after the guns were mechanized. The exnovation of existing knowledges, operating procedures, and structural poses can be more awkward than Figure 10.2 suggests. One of the major ways of their removal has been in the redeployment of staff – directly on to the labour market and indirectly in new assignments. For example, some organizations have removed whole plants and divisions at the operating level and have reformed their techno-structures around the new techno-economic paradigm (for example, Smith, Child, and Rowlinson 1990).

Exnovation may occur almost silently as when corporate ideologies are revised. Major changes in equipment, the built environment, and in raw materials were introduced during a ten-year period to transform the traditional manorial labour market of Cadbury's in Birmingham (Smith, Child, and Rowlinson forthcoming, 1990). In this case there was the exnovation of certain facets of the works council, of some plants, and of a significant segment of the techno-structure. These changes were preceded by the rescripting of the existing corporate ideology to reintro-duce much earlier themes of efficiency and to blend these with 'small is beautiful' campaign.

Chapter eleven

Innovation studies: technology, organization, and knowledge

Introduction

The development of an innovation perspective within organization studies has been central to the whole book. This final chapter summarizes the main themes and focuses on the relationships between technology, organization, and knowledge.

Review

Part I reported and examined the emergence of innovation studies as an area of enquiry which is shared amongst several academic disciplines: economics of innovation, manufacturing strategy, macro organization behaviour, industrial marketing, and the management of technology. Similarly, in corporations the problematic of innovation is leading to the recombination of existing functions into new areas such as 'strategic technologies'. Chapter 1 identified the analytic limitations of the orthodox mainstream to handle innovation. The emergence of innovation studies has mainly occurred through very significant detours from within existing disciplines, and so innovation studies is not yet established as the mainstream for those disciplines (for example, economics). Explicating innovation requires the replacement of the radical versus incremental distinction with a more discriminating framework. A fivefold typology of degrees of innovation was presented and the differences between entrenching and altering innovations were introduced. It was concluded that macro organization behaviour requires re-paradigming rather than the simplistic addition into efficiency-oriented texts of a few chapters on organizational transitions.

Chapter 2 presented the basic prospectus. Innovation studies was distinguished from efficiency studies and the time-free theorizing was distinguished, theorizing in which temporality was fully recognized. Innovation studies should combine prescriptive theorizing with descriptive quasi-theories which are multilevel and which examine the uneven, lumpy, deceptive unfolding of events – both external to organizations and also internal within organizations – with attention to the specificities of time and space. Innovation studies does not entirely abandon earlier law-like, time-free prescriptive frameworks, yet the development of innovation studies does require that the usage of such approaches includes a much more extensive descriptive analysis of contexts for both the academic and for the practitioner. One objective is to map situations so that contours and trends can be detected, and so that future scenarios enable intelligent, flexible planning. The dangers of the simplistic reliance on prescriptive frameworks for innovation are that some of these are confusing. Analytic priority should be given to the learning capabilities of the structural repertoires of organizations.

Part II concentrated upon clarifying the core constructs and the basic assumptions. The definition of innovation cannot be equated with equipment because innovation is a configuration which is dynamic and whose constitution is often an unintended outcome of the actions and interactions of many players and situations. Innovations emerge and are then subjected to the selective mechanisms of the environment. Many innovations disappear. So, although the directions of innovative activity are socially constructed, their survival and continual re-invention is shaped by the selection environment. Mechanisms of selection include: the economy (for example, sailing-ship effect), the institutional fabric, and specific resource dependencies.

Given these basic constructs and assumptions the analysis turned to the twin processsses of innovation-diffusion (Part III) and innovation-design (Part IV).

Part III focused on innovation-diffusion. Previous approaches to innovation have tended to underplay the role of the user in shaping innovations, yet the users role is especially significant with the process innovations which are at the foundations of contemporary economic competition. In chapter 5 attention was directed to the need for envisioning innovation as a dynamic array of diverse processes which irregularly create periods of major economic and social restructuring. It was argued that the contemporary period could be viewed from this perspective, and that the unfolding of several different yet interconnected waves of generic-altering innovations will lead to the virtual refounding of existing sectors (for example, automobile production; food and drink) and will therefore marginalize many long-established forms of expertise, especially in the large multinationals. These periods of transition provide challenges to existing organizations and opportunities for entry to the new organizations. However, because the larger users (for example, IBM, GM, ICI) are often crucially involved in the diffusion process, they may well possess the capacities for transition. The seminal case is the transformation of IBM from a supplier of mechanical and electro-mechanical office equipment into the supplier of advanced electronics. The survival of some large-scale enterprises and the disappearance of others, especially the large British national corporations in the food, drink, and tobacco sectors (for example, Imperial Tobacco, Rowntrees) provides an interesting situation for the organizational ecology debates. The notion of sector maturity, as developed in the early 1970s, argued the case for large volume based on the American model of technology driven movement down the high-volume-learning curve. The Utterback and Abernathy model was the major analytic vehicle for this position. The introduction of de-maturity of sectors to describe the collapse of American hegemony is too simplistic because there are many respects in which the Japanese have developed the best practices from the 1960s with their production knowledge about process innovations. The Japanese prised open the limits in American paradigms, especially the notion that core processes should be buffered (cf. Thompson 1967). Those developments were stimulated by, and intermingled with, the burgeoning of new technologies in information, in manufacturing, in bio-engineering, and in raw materials. The diffusion of innovations was examined through a close scrutiny of the centre-periphery model and the identification of a series of inherent problems because of its neglect of the role of the user in shaping innovations. The examination of the centre-periphery model of diffusion emphasized its remarkable qualities as a diagnostic framework whilst highlighting its limitations when used outside the narrow confines of the American agricultural extension agencies and similar situations. The centre-periphery model is best suited to the 'A' type of innovations. It is a supplier-oriented framework which was incorporated into the theories of consumer marketing in the era of mass marketing. It is a framework which has utility under certain conditions and as an exemplar, but now requires complementing by the development of diagnostic frameworks which explore the role of the users in the development of innovations (that is, the 'B' type) and which is sensitive to the complex process innovations of the current period. These themes of Part III were illustrated by contrasting the British and Japanese ability to use American innovations (in chapter 7). The macro-market dimension of different states (for example, Japan and the USA) was explored for its attributes as an entraining context for innovation. The use of international comparisons raised doubts about the claim that re-adaptation to specific markets is a clear sign of international survival. Consequently, insightful analysis of readaptation in the USA by Lawrence and Dyer (1983) should be treated with caution.

Part IV focused on the roles of the user and introduced some of the ingredients for an alternative framework to that of the centre periphery model. One very important ingredient is the networks and the division of labour over innovation between firms (see chapter 8). The

Innovation-design

development of networks between organizations is a complement
to the dismemberment of many existing vertically-integrated enterprises. Also, the process of developing networks is enhanced by the
new forms of equipment and information-based raw materials which
link chains of organizations in a value-adding chain. At the organizational level the constructs of structural repertoire and corporate expertise
were used to examine the roles of the technostructure in innovation
(chapter 9). Organizations develop specific knowledges which become
embedded into procedures and into the built environment to provide
a solid, slowly changing – possibly inertial – framework to ongoing
activities. Strategic innovation is very expensive and its comprehension
requires a highly-developed approach to design and development
as a set of practices in which the future is both anticipated and also
treated as the opportunity for continuous incremental innovation.
Organizations which develop sophisticated languages for contextual
analysis and which create rich structural repertoires possess the potential to treat complex innovations in an entrenching mode, thereby
minimizing some of the requirements for massive, highly visible,
restructuring and quantum leaps. Corporate cultures provide the grammar
and syntax – the cognitive and effective infrastructure – for innovation
as a designed process (chapter 10).

Technology, organization, and knowledge: beyond the limits

The limits

The analysis of the relations between technology and organization has
been beset with problems, some of which arise from the directions taken
in the mid-1960s. The archaeological approach of this book has surfaced
much of that inheritance and has provided frameworks and concepts
which should be applied to future analysis. We have argued that innovation studies should take a processual perspective in which structuration
occupies a central position because innovation occurs within a dynamic,
evolving and politicized network of channels and of players with varying
intentions – à filière. The essential texture of à filière cannot – at the
moment – be readily examined through the variance approach because
this is being applied in a blunt way which deflects attention away from
the more useful quasi descriptive theories (Mohr 1982). In order to move
away from the most restrictive limits of the analytic inheritance of
organization studies, and to reveal the limits, it is necessary in this final
section to prise open the potentials for the future. There are two steps
in the analysis: to replace the genealogical perspective to the
technology/organization debates; to highlight the role of knowledge and
learning in future research and theory building.

Buffering technological cores

To a very large degree the early studies on technology and organization equated technology with equipment, and so excluded the disembodied knowledge, the spatial forms, and the raw materials. These limits are evident in the four main streams of investigation: the sociotechnical perspective; in the many case studies of how new equipment altered earlier structures; in the contingency theory of Woodward; and in the Aston Programme and similar variance approaches. During the later 1960s there were two strong theoretical attempts to break away from those limitations and to provide a new sense of direction: by Thompson (1967) and by Perrow (1967). The contribution by Thompson tended to entrench established tendencies around a set of dangerous assumptions which gave efficiency priority over innovation, whilst the approach of Perrow has been only slowly used to alter established theory and to create a new perspective on innovation. Why were those four early streams limited and how did those limitations carry forward on to the perspectives of Thompson and Perrow?

The sociotechnical approach to technology included new forms of equipment in their spatio-temporal deployment (for example, long-wall mining machinery in Durham coal-mines) and also took into account the degree of uncertainty (that is, variance analysis) inherent in particular raw materials like the narrow, uneven, twisting British coal-seams. The perspective was highly visible during the 1960s, yet soon passed into almost total obscurity, in part because its theories were considered to be too closed to enable problem solving, and because many of its exponents preferred to promote a universal solution (for example, the autonomous work group) rather than a practice of diagnosis and policy making (Clark 1972a). The decline in interest was reinforced by the incompatibility between doing sociotechnical studies and maintaining the academic timetable and the forms of publishing which entrain most academics, especially in the USA. That said, it was in the USA that the most effective attempt at the diffusion of a diagnostic practice occurred from UCLA under the leadership of Louis E. Davis and colleagues. Certain features of the sociotechnical school deserved to be carried forward (for example, Miller and Rice 1967), but these were largely omitted from the abstract development of organization design theory (Clark 1975).

In many respects the plethora of case studies from this period on the implications of new equipment for organization structure and roles both sustained the image of a sociotechnical complex whilst otherwise failing to develop a quasi descriptive theory. For example, sociologists from the tradition of industrial sociology in Britain created a substantial portfolio of cases covering diverse sectors and showing quite substantial changes in work organization. However, those studies were atheoretical, simplistically longitudinal, rarely-examined failures, and did not compare the competitiveness of British innovations with what was happening in other countries. Neither did they develop a useful analysis of the design process and of the alternatives which might have been considered (cf. Miller and Rice 1967). The one serious attempt to synthesize across cases and to take into account the cultural contexts and structural conflicts was ignored (see Touraine 1965).

Woodward's contribution to the technology/organization issue was noted in the discussion of soft determinism in chapter 3. Her interpretation of the first large-scale investigation of how various facets of organization and of equipment was clustered into patterns was a major stimulus to the subsequent development of variance studies. Her approach could be translated into an abstract model (see Clark 1987 91 f.) and hence contributed to the development of a contingency perspective on technology and organization. However, the uneasy fusions between description and prescription served to obscure some of the principal contributions, and these complexities were enhanced rather than clarified by the attempt to incorporate the knowledge perspective (Woodward 1970).

The Aston Programme (for example, Pugh and Hickson 1976) serves as the clear example of a variance approach attempting to develop an empirical theory. The major impact was to highlight the variable of size – the confounded variable according to Weick (1979) – and to create the impression that equipment and its configuration had relatively few consequences for organization. These widely-accepted conclusions overpowered the earlier claims of the sociotechnical school, the case perspective, and the directions suggested by Woodward. Yet the scale for operations technology seems to be dubious (Clark 1987: 91 f.) and there are other problems (Child 1984). Yet subsequent studies often used the Aston scale of operations technology. The variance perspective often confused descriptions of what *is*, with the claims of prescriptive theory. There is little reason to assume that existing organizations are contingently fitted to their context, because some/many will disappear in the period after the research. For example, very few of the original fifty-two plants studied in Birmingham by the Aston Programme now exist – presumably they were not sufficiently innovative and were inefficient.

Three of the four streams of analysis of the technology/organization issue disappeared from view or were compacted into the main stream of variance studies. Considerable research has been undertaken within the variance format, yet successive reviews indicate very slight progress.

In the later 1960s Thompson and Perrow sought to reformulate the perspective on technology. Thompson's (1967) has been very influential in entrenching and so requires substantial revision. It is Perrow's perspective which provides the better analytic key. Thompson sought to derive a small number of key constructs and propositions to guide prescription about the best ways of organizing for technology. The important contributions included the distinctions between different kinds of interdependency *within* organizations (for example, pooled, reciprocal, sequential), and especially in the reasoning that management should seek to buffer their core technologies from the many turbulences arising in the external environment. The notion of buffering the core technology became an unexamined wooden horse in American management paradigms. The significance of buffering is that it is a just-in-case world view and this may be seen very clearly by recalling Abernathy's (1978) seminal study of the preference by Ford for efficiency rather than innovation. Buffering provided a sense of control to the professionalized technostructure of American corporations because it seemingly facilitated the standardization of corporate knowledge about production (see Larsen 1977). Buffering had other less desirable longer-term consequences for innovation, and so it is relevant to note that Abernathy's analysis was heavily dependent on the interpretation of knowledge technology (Perrow 1967; Larsen 1977: 197 f.).

Technology and organization as knowledge and learning

Knowledge and organizational learning have become major features in the examination of technological change and organizational change. The economics of knowledge, for example, is now established, and includes issues such as learning (for example, by doing, by using), the appropriation of economic benefits from knowledge development, the embodiment of knowledge and the role of problem-solving capabilities (Nelson and Winter 1982; Dosi 1984). The economics of learning capabilities is central to studies of David (1975) and to Rosenberg (1982). These contributions have extended the scope of analysis from the (buffered) organization into interfirm networks (see chapter 8), and into the role of the division of labour amongst chains and sectors of firms in the production of knowledge. Moreover, the focus on knowledge has extended process theories from their attention to the firm-in-sector (Whipp and Clark 1986: chapter 2) on to and into the societal filière of institutions and the consequences of those social bases for industrial innovation (Maurice, Sellier, and Sylvestre 1986; Clark 1987). The institutional filière deserves much more attention.

There is considerable potential for developing and extending the position taken on technology and organization learning. The extension should include the careful review of what elements can be usefully carried forward from the earlier sociotechnical perspective (for example, the spatio-temporal configurations and their oscillations). Perrow focused on knowledge technology as the central feature of a new theory by identifying the informational interactions between the inherent demands

215

of raw materials (as defined by observers and as perceived by the operative) on both operators and their usage of equipment. So any changes which increase or decrease the extent of standardization in the knowledge utilized by particular groups are likely to have considerable implications (Larsen 1977: 197 f.). Hence the firm should be examined for its repertoires of problem-building knowledges, and to explore the limits of corporate stocks of knowledge to advance existing levels of problem solving. New forms of equipment and raw material such as network technologies both 'require' new forms of knowledge and also enable their constitution (see Figure 3.1). As yet there are few empirical studies which test and develop the insights which the theoretical exposition of organizational knowledge and learning have suggested. Nor are there established practices to critique and develop the firm specific knowledge, although there are interesting developments in the soft modelling approaches to organization which might be adapted to give a more serious thrust to the problem of innovation (for example, Checkland 1981). It must be emphasized that such approaches have to be applied at two levels:

(a) the societal filière of institutions and culturally specific predispositions;
(b) the firm in its sectoral contexts because innovation – from incremental to radical – requires agentic actions (and luck!) by a combination of players in the construction of a usable theory of innovation relevant to organizations.

The stress has been given to those aspects of technology and organization which, although central, were under-developed in earlier theorizing, research and policy making. Innovation studies is a perspective on technology and organization: a world view. We have aimed to highlight the relevance of an innovation world view, to provide a prospectus and to show why its main constructs and insights should be incorporated into organization studies.

Bibliography

Abernathy, W.J. (1978) *The Productivity Dilemma Roadblock to Innovation in the Automobile Industry*, London: Johns Hopkins University.

Abernathy, W.J. and Clark, K.B. (1985) 'Innovation: mapping the winds of creative destruction', *Research Policy* 14: 3–22.

Aggarwal, S.C. (1985) 'MRP, JIT, OPT, FMS: making sense of production operating systems', *Harvard Business Review*, September/October.

Alderson, W. (1965) *Dynamic Marketing Behaviour. A Functionalist Theory of Marketing*, Homewood: Irwin.

Aldrich, H.E. (1979) *Organisations and Environments*, New Jersey: Prentice Hall.

Arndt, J. (1983) 'The political economy paradigm: foundation for theory building in marketing', *Journal of Marketing* Fall, 44–54.

Barley, S.R. (1986) 'Technology as an occasion for structuring: evidence from observations of CT, scanners of the social order of radiological departments', *Administrative Science Quarterly*, 31: 78–108.

Bennis, W., Benne, K., and Chin, R. (eds) (1961 and 1985) *The Planning of Change Readings in the Applied Social Sciences*, New York: Holt, Rinehart & Winston.

Blau, P.M. (1968) 'The hierarchy of authority in organisations', *American Journal of Sociology* 73: 453–67.

Blau, P.M. and Scott, W.R., (1961) *Formal Organizations*, London: Routledge.

Boisard, P. and Etablier, M.-T. (1987) 'Le camembert: Normand or norme deux modeles de production', *L'Industrie Fromagère*, Cahiers de Centre d'Etude et d'Emploi, 30, Paris: Presses Universitaires de France.

Boisot, M. and Child, J. (1987) 'Efficiency, ideology and tradition in the choice of transactions governance structures – the case of China as a modernising society', Work Organization Research Centre (WORC) Working Paper, no. 21, Aston Business School.

Braudel, F. (1972) *The Mediterranean & the Mediterranean World*, London: Collins.

Bright, J.R. (1958) *Automation of Management*, Boston: Irwin.

Bristor, J.M. and Ryan, M.J. (1987) 'The buying centre is dead: long live the buying centre', in Wallendorf, M. and Anderson, P.F., *Advances in Consumer Research*, Provo Utah: Association for Consumer Research, pp. 255–8.

217

Brooks, H. and Kelly, M.E. (1986) 'Technical, economic and organisational factors explaining the diffusion of programmable automation of machining', Working Paper J.F. Kennedy School of Government, Harvard University.

Brown, L.A. (1981) *Innovation Diffusion. A New Perspective*, London: Methuen.

Burawoy, M. (1984) *The Politics of Work*, London: Heinemann.

Burns, T. and Stalker, G.M. (1961) *The Management of Innovation*, London: Tavistock.

Burrell, G. and Morgan, G. (1979) *Sociological Paradigms and Organizational Analysis*, London: Heinemann.

Chandler, A.D. (1977) *The Visible Hand. The Managerial Revolution in America*, Cambridge: Bellknap.

Checkland, P. (1981) *Systems Thinking, Systems Practice*, New York: Wiley.

Child, J. (1969) *British Management Thought*, London: Allen & Unwin.

Child, J. (1984) *Organization: a guide to problems and practice*, London: Harper Row.

Child, J. (1987) 'Information-Technology, Organisation and the Response to Strategic Challenges', *California Management Review*.

Clark, I.M. (1985) *The Spatial Organization of Multinational Corporations*, London: Croom Helm.

Clark, K.B. (1985) 'The interaction of design hierarchies and market concepts in technological evolution', *Research Policy* 14: 235–51.

Clark, N.G. (1985) *The Political Economy of Science and Technology*, Oxford: Blackwell.

Clark, P.A. (1972a) *Organisational Design: theory and practice*, London: Tavistock.

Clark, P.A. (1972b) *Action Research and Organisation Change*, London: Harper Row.

Clark, P.A. (1975) 'Key problems in organisation design', *Administration and Society* 7(2): 213–56.

Clark, P.A. (1976) 'Some analytic requirements of an applied organization science', in Kilman, R., Slevin, D., and Pondy, L. (eds) *Managing Organizational Design*, New York: Elseview.

Clark, P.A. (1985) 'A review of theories of time and structure for organisational sociology', in Bachrach, S.B. and Mitchell, S.M. (eds) *Research in the Sociology of Organisations*, vol. 4, Connecticut: JAI Press.

Clark, P.A. (1986) 'The economy of time and the managerial division of labour in the British construction industry', *Proceedings of Bartlett International Summer School* 7, London: University College.

Clark, P.A. (1987) *Anglo-American Innovation*, New York: de Gruyter.

Clark, P.A. (1989) 'Organisational analysis and organisational chronologies', in Hassard, J. and Pymm, D. *The Theory and Philosophy of Organisations: critical issues and new perspectives*, London: Croom Helm.

Clark, P.A. and DeBresson C. (1987, 1988) 'Organization transitions and sector technology life cycle models applied to car firms: Rover (1896–1987)', *Proceedings ASAC*, Toronto University.

Clark, P.A. and De Bresson, C. (1989) 'Innovation-design and innovation poles', in Loveridge, R.L. and Pitt, M. (ed.) *Strategic Management of Technological Innovation*, London: Wiley.Clark, P.A. and Starkey, K.P. (1988) *Organization Transitions and Innovation-design*, London: Frances Pinter.Clark, P.A. and Windebank, J. (1985) 'Innovation and Renault 1900–1982 Product, Process and Work Organisation', WORC Working Paper, no. 13, Aston Business School.

Cohen, M.D., March, J.G., and Olsen, J.P. (1972) 'A garbage can model of organisation choice', *Administrative Science Quarterly* 17: 1–25.

Coombs, R., Saviotti, P., and Walsh, V. (1987) *Economics and Technical Change*, London: Macmillan.

Cork, D. (1985) *The Guide to Computer Aided Production Management*, London Institute of Production Engineers.

Crozier, M. (1964) *The Bureaucratic Phenomenon*, London: Tavistock.

Crozier, M. (1973) *The Stalled Society*, New York: Viking Press.

Crozier, M. (1984) *The Trouble with America*, Berkeley: University of California Press.

Cusumano, M.A. (1985) *The Japanese Automobile Industry. Technology and Management at Nissan and Toyota*, Boston: Harvard University Press.

Cyert, R.M. and March, J.G. (1963) *A Behavioural Theory of the Firm*, New Jersey: Prentice Hall.

Damanpour, F. (1987) 'Organization adoption of innovations: a review of research findings on three dimensions', Graduate School of Management, New Jersey: Rutgers University.

David, P.A. (1975) *Technological Choice, Innovations and Economic Growth. Essays on American British Experience in the 19th Century*, Cambridge: Cambridge University Press.

David, P.A. (1986) 'Understanding the economics of QWERTY: the Necessity of History, in Parker, W.N. (ed.) *Economic History and the Modern Economist*, Oxford: Blackwell.

David, P.A. and Bunn, J.A. (1987) ' "The battle of the systems" and the evolutionary dynamics of network technology rivalries', Working Paper, no. 14, Center for Economic Policy Research, Stanford University.

Davies, S.R. (1979) *The Diffusion of Process Innovations*, Cambridge: Cambridge University Press.

DeBresson, C. and Murray, B. (1982) *Innovation in Canada*, Cooperation Research Unit on Science & Technology, British Columbia.

Dore, R.P. (1973) *British Factory, Japanese Factory*, London: Allen & Unwin.

Dosi, G. (1982) 'Technological paradigms and technological imperatives', *Research Policy* 11(3): 147–62.

Dosi, G. (1984) *Technical Change and Industrial Transformation*, London: Macmillan.

Dow, G.K. (1988) 'Configurational and coactivational views of organization structure', *Academy of Management Review* 13(1): 53–64.

Dumas, A. (1988) 'Silent designers', in Clark, P.A. and Starkey, K. *Organization Transitions and innovation-design*, London: Frances Pinter.

Eisenstadt, S.N. (1973) 'Some reflections on the "crisis" in sociology',

Sociologische Grds 4: 255–69.

Emery, F.E. (1959) 'Characteristics of socio-technical systems: a critical review of theory and facts', Tavistock Institute, TIHR527.

Etzioni, A. (1961) *A Comparative Analysis of Complex Organizations*, New York: Free Press.

Evan, W.H. (1966) 'The organizational set: toward a theory of interorganizational relations', in Thompson, J.D., *Approaches to Organizational Design*, Pittsburgh: University of Pennsylvania Press.

Eveland, J.D. (1981) 'Measuring the Innovation Process in Public Organisations', Working Paper, Washington: National Science Foundation.

Eveland, J.D. and Rogers, E.M. (1980) 'Measuring the innovation process in public organisations', Working Paper, Washington: National Science Foundation.

Foucault, M. (1977) *Discipline and Punish. The Birth of the Prison*, London: Penguin.

Foxall, G.R. (1984) *Corporate Innovation. Marketing and Strategy*, London: Croom Helm.

Foxall, G.R. (1987) 'Markets, hierarchies and user-initiated innovation', Working Paper, Strathclyde University.

Freeman, C. (1974) *The Economics of Industrial Innovation* (2nd edn) London: Frances Pinter.

Freeman, C. (1981) 'The diffusion of innovations – microelectronic technology', in Lundvall, B. and Christensen, P.R. (eds) *Technology and Employment*, Alborg: Alborg University Press.

Freeman, C. (1982) *Unemployment and Technical Innovation*, London: Frances Pinter.

Freeman, C. (1983) (ed.) *Longwaves in the World Economy*, London: Butterworth.

Freeman, C. (1986) 'Induced innovation, diffusion of innovations and business cycles', Edinburgh: Canadian Studies Conference.

Galbraith, J.R. (1977) *Organizational Design*, Reading, Mass.: Addison Wesley.

Gallie, D. (1978) *In Search of the New Working Class: automation and integration within the capitalist enterprise*, Cambridge: Cambridge University Press.

Gearing, E. (1958) 'The structural poses of the eighteenth century Cherokee villages', *American Anthropologist* 60: 148–57.

Ghoshal, S. and Bartlett, C.A. (1987) 'Organising for innovations: case of multinational corporation', Working Paper 87/04, Harvard Business School.

Giddens, A. (1979) *Central Problems in Social Theory: action, structure and contradiction in social analysis*, London: Macmillan.

Giddens, A. (1985) *The Constitution of Society*, Oxford: Polity Press.

Gille, B. (1978) *Histoire des Techniques*, Paris: Pleiade.

Goffman, E. (1974) *Frame Analysis*, New York: Harper Row.

Gold, B. (1981) *Evaluating Technological Innovations, Methods, Expectations and Findings*, Lexington: Lexington Press.

Griliches, Z. (1957) 'Hybrid corn: an exploration in the economics of technological change', *Econometrica* 25(4): 501–22.

Hagerstrand, T. (1952) 'The propagation of innovation waves', *Human Geography* 4: 3–19.

Håkansson, H. (1986) 'Industrial technical development: a network approach', draft manuscript, University of Stockholm.

Halberstram, D. (1986) *The Reckoning*, New York: Avon.

Haley, A. (1972) *Wheels*, New York: Random House.

Hall, R.I. (1984) 'The natural logic of management policy making: its implications for the survival of an organisation', *Management Science* 30(8): 905–27.

Harley, C.K. (1971) 'The shift from sailing ships to steamships: a study in technological change and its diffusion', in McCloskey (ed.) *Essays on a Mature Economy: Britain after 1840*, London: Methuen.

Hayes, R.H. and Wheelwright, S. (1985) *Restoring our Competitive Edge, Competing through Manufacturing*, New York: Wiley.

Hickson, D.J., *et al.* (1986) *Top Decisions. Strategic Decision Making in Organizations*, Oxford: Blackwell.

Hoskin, K.W. and Macve, R. (1988) 'The genesis of accountability: the West Point connections, *Accounting, Organizations and Society* 13: 37–73.

Hounshell, D.A. (1984) *From the American System of Mass Production, 1800–1932: the development of manufacturing technology in the United States*, Baltimore: Johns Hopkins University.

Hughes, T. (1983) *Networks of Power: electrification in western society 1880–1930*, Baltimore: Johns Hopkins University.

Hull, F.M. and Azumi, K. (1987) 'Invention rates and R & D in Japanese factories', Working Paper, Rutgers University, New Jersey.

Johanson, J. and Mattson, L.C. (1987) 'Interorganisational relations in industrial systems: a network approach compared with the transaction cost approach', *International Studies of Management and Organisation* 17(1): 34–48.

Johnson, T.H. and Kaplan, R.S. (1987) *Relevance Lost: The Rise and Fall of Management Accounting*, Boston: Harvard Business School Press.

Joyce, J. (1960) *Ulysses*, London: Bodley Head.

Kanter, R.M. (1984) *The Change Masters, Corporate Entrepreneurs at Work*, London: Allen & Unwin.

Kantrow, A.M. (1980) 'The strategy technology connection', *Harvard Business Review*, July/August: 6–21.

Kaplinsky, R. (1984) *Automation: the technology and society*, London: Longman.

Kaplinksy, R. (1986) Electronics-Based Automatic Technologies and the onset of Manufacture, Working Paper. Institute of Development Studies, Sussex University.

Karpik, L. (1978) 'Organisations, institutions and history', in L. Karpik (ed.) *Organization and Environment: Theory Issues and Reality*, Beverley Hills: Sage.

Kennedy, P. (1987) *The Rise and Fall of Great Powers: economic change and military conflict 1500 to 2000*, New York: Random House.

Kimberly, J.R. (1981) 'Managerial innovation', in Nystrom, P.C. and Starbuck, W.H. *Handbook of Organizational Design*, vol. I, Oxford: Oxford University Press.

Kinch, N. (1984) 'The longterm development of a supplier-buyer relationship: the case of the Olofstrom-Volvo relationship', Upsalla University, Sweden: Centre for International Business Studies.

Kochar, A.K. (1988) 'Design and operation of manufacturing systems in Japanese industry'. Report, England: Bradford University.

Kondratiev, N.D. (1976) 'The long waves in economic life', *Lloyds Bank Review*, pp. 41–60.

Lammers, C.J. (1981) 'Contributions to organization sociology: Part 1: Contributions to sociology – a liberal view', *Organization Studies* 2/3: 267–86.

Landes, D. (1969) *The Unbound Prometheus: technological change and industrial development in western Europe 1750 to the present*, London: Cambridge University Press.

Larsen, M.S. (1977) *The Rise of professionalism: A sociological analysis*, Los Angeles: University of California Press.

Latour, B. (1986) 'Machiavelli and the role of engineers', Working Paper, Edinburgh: Edinburgh University.

Latour, B. (1988) *The Pasteurization in France*, Cambridge, Mass: Harvard University Press.

Lawrence, P.R. and Dyer, R. (1983) *Renewing American Industry*, New York: Free Press.

Lawrence, P.R. and Lorsch, J.D. (1967) *Organization and Environment* Cambridge, Mass.: University of Harvard Press.

Leavitt, H.J. and Whisler, T.L. (1958) 'Management in the 1980s', *Harvard Business Review* November/December: 41–51.

Leonard-Barton, D. (1987) 'Implementing new technology: the transfer from developers to operations'. Working Paper, Harvard Business School/Research Division.

Lewin, A.Y. and Minton, J.W. (1986) 'Determining organizational effectiveness, another look and an agenda for research', *Management Science* 32(5): 514–38.

Litwak, E. (1961) 'Models of bureaucracy which permit conflict', *American Journal of Sociology* 67: 177–84.

Lord, R.G. (1988) 'Scripts as determinants of purposeful behaviour in organization', *Academy of Management Review* 2: 265–77.

Love, J.F. (1986) *McDonalds: behind the arches*, New York: Bantam Press.

McDonald, S., Lamberton, D., and Mandeville, T. (eds) (1983) *The Trouble with Technology*, London: Pinter.

McIntosh, N. (1985) *Social Software of Accounting*, London: Wiley.

Mackrimman, K.R. and Wagner, C. (1986) 'Expert Systems and Creativity', Working Paper 1209, University of British Columbia Press.

Mandel, E. (1978) *Late Capitalism*, London: Verso.

Mansfield, G. (1958) *The Economics of Technological Change*, New York: Norton.

March, J.G. and Simon, H.A. (1958) *Organisations*, New York: Wiley.

Maurice, M., Sellier, F., and Sylvestre, J.-J. (1986) *The Social Foundations of Industrial Power. A Comparison of France and Germany*, Cambridge, Mass.: MIT Press.

Mensch, G. (1979) *Stalemates in Technology: innovations overcome the depression*, Massachusetts: Ballinger.

Merton, R.K. (1957) *Social Theory and Social Structure* New York: Free Press.

Metcalfe, J.S. (1981) 'Impulse and diffusion in the strategy of technological change', *Futures* 13(5) 347–59.

Miles, R.H. (1980) *Macro Organizational Behavior*, Glenview: Scott, Foresman & Co.

Miles, R.E. and Snow, C.C. (1978) *Organizational Strategy, Structure and Process*, New York: McGraw Hill.

Miles, R.E. and Snow, C.C. (1986) 'Organizational: new concepts for new forms', *California Management Review* 28(3): 62–73.

Millard, R.I. (1985) 'MRP is none of the above', *Production and Inventory Management*, first quarter: 22–9.

Miller, D. and Friesen, P.H. (1984) *Organisations. A Quantum View*, New Jersey: Prentice-Hall.

Miller, E.J. and Rice, A.K. (1967) *Systems of Organisation*, London: Tavistock.

Mintzberg, H. (1983) *Structures in Fives: designing effective organizations*, Englewood Cliffs, NJ: Prentice Hall.

Mintzberg, H., Ralsinghani, D., and Thoret, A. (1976) 'The structure of unstructured design processes', *Administrative Science Quarterly* 21: 246–75.

Mohr, L.B. (1982) *Explaining Organizational Behavior: the limits and possibilities of theory and research*, San Francisco: Jossey-Bass.

More, R.A. (1986) 'Developer/adopter relationship in new industrial product situations', *Journal of Business Research* 14 (June): 501–17.

Mowery, D.C. and Rosenberg, N. (1979) 'The influence of market demand upon innovation: a critical review of some empirical studies', *Research Policy* 8: 103–53.

Nelson, D. (1983) 'Taylorism in American society 1900–1930', in Montmollin, M. and Pastre, O. (eds) *Le Taylorism*, Paris: La Decomete.

Nelson, R. and Winter, S. (1977) 'In search of a useful theory of innovation', *Research Policy* 6(1): 36–77.

Nelson, R. and Winter, S. (1982) *An Evolutionary Theory of Economic Change*, Cambridge, Mass.: Harvard University Press.

Noble, D. (1977) *America by Design, Science, Technology and the Rise of Corporate Capitalism*, New York: Knopf.

Noble, D. (1984) *Forces of Production. A Social History of Industrial Automation*, New York: Knopf.

Pavitt, K. (1984) 'Sectoral patterns of technical change: towards a taxonomy and a theory', *Research Policy* 13: 343–73.

Pavitt, K. (1987) 'The objectives of technology policy', *Science and Public Policy* 14(4): 182–88.

Penrose, E.T. (1959) *The Theory of the Growth of the Firm*, Oxford: Blackwell.

Perez, C. (1983) 'Structural change and assimilation of new technologies in the economic and social systems', *Futures* October: 357–75.

Perrin, (1977) *Engineering: Terminology & Economic Functions*, Paris: OECD.

Perrow, C. (1967) 'A framework for the comparative analysis of organizations', *American Sociological Review* 32: 194–208.

Peters, T.J. and Waterman, R.H. (1982) *In Search of Excellence*, New York: Harper Row

Pettigrew, A. (1973) *The Politics of Organizational Decision-making*, London: Tavistock.

Pettigrew, A. (1985) *The Awakening Giant Continuity and Change at ICI*, Oxford: Blackwell.

Pinch, T.J. and Bijker, W.E. (1987) 'The social construction of facts & artefacts: or how the sociology of science & the sociology of technology might benefit each other', in Bijker, W.E., Hughes, T.P., and Pinch, T. 1987) *The Social Construction of Technological Systems* London: MIT Press.

Piore, M. and Sabel, C.A. (1984) *The Second Industrial Divide: Possibilities for Prosperity*, New York: Basic Books.

Porter, M.E. (1983a) *Competitive Strategy: Techniques for Analysing Countries and Competitors*, New York: Free Press.

Porter, M.E. (1983b) 'The technological dimension of competitive advantage', in Rosenbloom, R.S. (ed.) *Technological Innovation, Management and Policy*, vol. 1, CT JAI Press.

Porter, M.E. (1985) *Competitive Advantage: Creating and Sustaining Superior Performance*, New York: Free Press.

Prunty, J.P., Smith, L.M., Dwyer, D.C., and Kleine, P.F. (1987) *The Fate of an Innovative School*, London: Falmer Press.

Pugh, D.S. and Hickson, D.J. (1976) *Aston Programme: Volume One*, London: Saxon House.

Pulos, A.J. (1983) *American Design Ethic. A History of Industrial Design to 1940*, Boston: MI Press.

Quinn, J.B. (1980) *Strategies for Change – Logical Incrementalism*, New York: Irwin.

Quinn, J.B., Mintzberg, H., and James, R.M. (1988) *The Strategy Process: concepts, contexts and cases*, Englewood Cliffs, NJ: Prentice Hall.

Raynor, R. (1987) 'MRPII: is there a payoff?' *Industrial Computing*, November: 21–5.

Reich, R.B. and Donahue, J.D. (1985) *New Deal: the Chrysler revival and the American system*, New York: Basic Books.

Rhodes, E. and Wield, D. (eds) (1985) *Implementing New Technologies: choice, design and change in manufacturing*, Oxford: Blackwell.

Rogers, E.M. (1962, 1983, 3rd edn) *Diffusion of Innovations*, New York: Free Press.

Rogers, E.M. (1986) *Communications Technology: the new media in society*, New York: Free Press.

Rogers, E.M. (1987) 'Progress, problems and prospects for network research: investigating relationship in the age of electronic communication technologies', Sun Belt Networks Conference VII, Florida, USA.

Rogers, E.M. and Rogers, R.K. (1976) *Communication in Organisation*, New York: Free Press.

Rogers, E.M. and Shoemaker, F.S. (1971) *Communication of Innovations: a cross-cultural perspective*, New York: Free Press.

Rosenberg, N. (1969) *The American System of Manufacturing*, Edinburgh: Edinburgh University Press.

Rosenberg, N. (1976, 1982) *Perspectives on Technology*, Cambridge: Cambridge University Press.

Rosenberg, N. (1982) *Inside the Black Box. Technology and Economics*, New York: Cambridge University Press.

Rosenberg, N. and Steinmueller, E.W. (1988) 'Can Americans learn to become better imitators?' Working Paper, Stanford University.

Rothwell, R. (1986) 'Innovation and re-innovation: a role for the user', *Journal of Marketing Management* 2: 109–23.

Rothwell, R. and Gardiner, P. (1983) 'The role of design in product and process change', *Design Studies* 4(3): 161–9.

Rothwell, R. and Gardiner, P. (1985) 'Invention, innovation, re-innovation and the role of the user: a case study of British Hovercraft development', *Technovation* March: 167–86.

Roy, R. and Wield, D. (eds) (1986) *Product Design and Technological Innovation*, Milton Keynes: Open University Press.

Ruel, S. (1987) *Japanese Financial Institutions in Britain*, no publisher.

Ryan, B. and Gross, N.C. (1943) 'The diffusion of hybrid seed corn in two Iowa communities', *Rural Sociology* 8: 15–24.

Sabel, C.A. (1988) 'Re-regionalization of production', paper delivered in Paris at Logiques d'Enterprise et Formes de Legitomite, Colloque International.

Sahal, D. (1981) *Patterns of Technological Innovation*, New York: Addison Wesley.

Schank, R.C. (1977) *Scripts, Plans, Goals and Understanding: an inquiry into human knowledge*, New York: Wiley.

Schon, D.A. (1971) *Beyond the Stable State*, New York: Randam House.

Schonberger, R.J. (1982) *Japanese Manufacturing Techniques. Nine Hidden Lessons in Simplicity*, New York: Free Press.

Schumpeter, J. (1939) *Business Cycles*, New York: Oxford University Press.

Scott-Kemmis, D. (1983) 'The nature and significance of incremental technical change: a case-study of a U.K. paper mill', Paper prepared for the Technical Change Centre.

Smelser, N.J. (1976) *Comparative Methods in the Social Sciences*, Englewood Cliffs, NJ: Prentice-Hall.

Smith, C., Child, J., and Rowlinson, M. (forthcoming 1990) *Innovation in Work Organisation – the Cadbury Experience*, Cambridge: Cambridge University Press.

Sobel, R. (1984) *Car Wars*, New York: Dutton.

Stoneman, P. (1976, 1983) *Technological Diffusion and the Computer Revolution*, Cambridge: Cambridge University Press.

Teece, D.J. (1987) 'Profiting from technological innovation: implications for integration collaboration, licensing and public policy', *Research Policy* 5(6): 285–305.

Thevenot, L. and Boltanski, L. (1988) *Economies de Grandeur*, Paris: Presses Universitaires de France.

Thompson, E.P. (1967) 'Time, work discipline and industrial capitalism', *Past & Present* 38: 56–97.

Thompson, J.D. (1967) *Organizations in Action*, New York: Wiley.

Thorelli, H. (1986) 'Networks: between markets and hierarchies', *Strategic Management Journal* 7: 37–51.

Bibliography

Touraine, A. (ed.) (1965) *Workers' Attitudes to Technological Change*, Paris: OECD.

Tushman, M.L. and Anderson, P. (1986) 'Technological discontinuities and organizational environment', *Administrative Science Quarterly* 31 March: 439–65.

Utterback, J.M. and Abernathy, W.J. (1975) 'A dynamic model of process and product innovation, *Omega* 3(6): 630–56.

von Hippell, E. (1982a) 'The appropriability of innovation benefit as a predictor of the source of an innovation', *Research Policy* 11: 95–115.

von Hippell, E. (1982b) 'The customer-active paradigm for industrial product generation', *Research Policy* 240–66.

von Hippell, E. (1983) 'Novel Product Concepts from Lead Users: segmenting users by experience', Working Paper no. 1476–83, Cambridge, Mass.: MIT (Massachusetts Institute of Technology).

Weick, K. (1969) *The Social Psychology of Organising*, Reading, Mass.: Addison-Wesley.

Whipp, R. and Clark, P.A. (1986) *Innovation and the Auto Industry. Product, Process & Work Organization*, London: Frances Pinter.

Whiteside, D. and Ambrose, J. (1984) 'Unsnarling industrial production: why top management is starting to care', *International Management* March.

Woodward, J. (1958) *Management and Technology*, London: HMSO.

Woodward, J. (1965) *Industrial Organization: Theory and Practice*, London: Oxford University Press.

Woodward, J. (ed.) (1970) *Industrial Organization: Behaviour and Control*, London: Oxford University Press.

Yates, B. (1983) *The Decline and Fall of the American Automobile Industry*, New York: Empire Books.

Yin, R.W. (1979) *Changing Urban Bureaucracies: how new practices become routinized*, Santa Monica: Rand.

Zaltman, G. and Duncan, R.B. (1977) *Strategies for Planned Change*, New York: Wiley.

Index